100 Multiplication Facts

I0427490

Our New Fact Is 2

2 Facts			Draw dashed lines from the facts to the correct answers!	
2 x 0 = 0		2 x 3		10
2 x 1 = 2		2 x 9		8
2 x 2 = 4		2 x 2		20
2 x 3 = 6		2 x 7		6
2 x 4 = 8		2 x 6		0
2 x 5 = 10		2 x 1		16
2 x 6 = 12		2 x 10		12
2 x 7 = 14		2 x 8		4
2 x 8 = 16		2 x 0		14
2 x 9 = 18		2 x 5		18
2 x 10 = 20		2 x 4		2

Write Your 3 Facts

Trace it	Answer it	Fill in the blanks	Fill in the blanks	Write the fact
3 x 6 = 18	3 x 6 =	x 6 =	x =	
3 x 2 = 6	3 x 2 =	3 x =	x =	
3 x 7 = 14	3 x 7 =	3 x =	x =	
3 x 1 = 3	3 x 1 =	3 x =	x =	
3 x 5 = 15	3 x 5 =	x 5 =	x =	
3 x 8 = 24	3 x 8 =	x 8 =	x =	
3 x 4 = 12	3 x 4 =	3 x =	x =	
3 x 0 = 0	3 x 0 =	3 x =	x =	
3 x 9 = 27	3 x 9 =	x 9 =	x =	
3 x 3 = 9	3 x 3 =	3 x =	x =	
3 x 10 = 30	3 x 10 =	3 x =	x =	

Let's Practice!

Let's make sure you have all your facts down! Answer each multiplication problem.

$$\begin{array}{r} 4 \\ \times\ 5 \\ \hline \end{array} \quad \begin{array}{r} 4 \\ \times\ 1 \\ \hline \end{array} \quad \begin{array}{r} 4 \\ \times\ 8 \\ \hline \end{array}$$

$$\begin{array}{r} 4 \\ \times\ 6 \\ \hline \end{array} \quad \begin{array}{r} 4 \\ \times\ 2 \\ \hline \end{array}$$

$$\begin{array}{r} 4 \\ \times\ 9 \\ \hline \end{array} \quad \begin{array}{r} 4 \\ \times\ 4 \\ \hline \end{array}$$

$$\begin{array}{r} 4 \\ \times\ 0 \\ \hline \end{array} \quad \begin{array}{r} 4 \\ \times\ 3 \\ \hline \end{array}$$

$$\begin{array}{r} 4 \\ \times\ 10 \\ \hline \end{array} \quad \begin{array}{r} 4 \\ \times\ 7 \\ \hline \end{array}$$

I can multiply by 5s: Top AND Bottom

Name: _____ Date: _____

Goal: ___ problems in ___ seconds/minutes

5 × 5	5 × 0	5 × 6	5 × 5	8 × 5	5 × 8
5 × 9	5 × 1	2 × 5	8 × 5	5 × 7	5 × 9
10 × 5	8 × 5	5 × 9	3 × 5	5 × 7	9 × 5
1 × 5	0 × 5	3 × 5	5 × 2	5 × 5	5 × 7
5 × 6	5 × 1	5 × 0	10 × 5	5 × 8	3 × 5
2 × 5	4 × 5	5 × 4	7 × 5	5 × 6	8 × 5
5 × 2	5 × 7	0 × 5	5 × 2	7 × 5	5 × 9
5 × 8	6 × 5	3 × 5	5 × 9	5 × 4	3 × 5
1 × 5	5 × 7	5 × 9	3 × 5	2 × 5	5 × 7
5 × 5	6 × 5	5 × 9	10 × 5	5 × 3	5 × 9

1

Contact the author:
coaches@resourceteachersguide.com

Table of Contents

Introduction

This multiplication facts practice book is designed to be a go-at-your-own-pace approach. We've broken the facts down into smaller steps with regular review to help take out the overwhelm that students may feel.

Our unique approach in using the small steps along with extra practice working with the facts and the answers provide students the best path to actual mastery of their facts.

Instructions For Use

Each lesson is broken down as follows:
1. Introduce the new fact with its answers.
2. Practice through writing and repetition
3. Apply what they've learned
4. Timing with fact on top or bottom
5. Timing with fact on top and bottom
6. Review Timings (mix of all the facts they've learned up to that point)

Before each timing sit down with your child and set a goal that will help you both know that the child has mastered that fact and can move on.

For our timed tests we recommend setting goals as follows:
- *Typical goal:* **60 problems in 60-90 seconds**
- *Students who struggle with processing or test anxiety:* **60 problems in 2-3 minutes**

Use the tracking sheet on page 5 to increase motivation and track progress.

TIP: If you feel you need more practice rip out the timing, put it in a sheet protector and use a dry erase marker!

Passing Multiplication

0	1	0-1	2
0-2	3	0-3	4
0-4	5	0-5	6
0-6	7	0-7	8
0-8	9	0-9	10

0-10	When I pass I get:

Our New Fact Is

0 Facts
0 x 0 = 0
0 x 1 = 0
0 x 2 = 0
0 x 3 = 0
0 x 4 = 0
0 x 5 = 0
0 x 6 = 0
0 x 7 = 0
0 x 8 = 0
0 x 9 = 0
0 x 10 = 0

Draw squiggly lines from the facts to the correct answers!

0 x 0	0
0 x 4	0
0 x 10	0
0 x 3	0
0 x 2	0
0 x 9	0
0 x 1	0
0 x 7	0
0 x 8	0
0 x 1	0
0 x 6	0

Write Your 0 Facts

Trace it	Answer it	Fill in the blanks	Fill in the blanks	Write the fact
0 x 3 = 0	0 x 3 =	0 x =	x =	
0 x 1 = 0	0 x 1 =	0 x =	x =	
0 x 8 = 0	0 x 8 =	0 x =	x =	
0 x 0 = 0	0 x 0 =	x 0 =	x =	
0 x 4 = 0	0 x 4 =	x 4 =	x =	
0 x 2 = 0	0 x 2 =	x 2 =	x =	
0 x 10 = 0	0 x 10 =	0 x =	x =	
0 x 7 = 0	0 x 7 =	0 x =	x =	
0 x 5 = 0	0 x 5 =	x 5 =	x =	
0 x 9 = 0	0 x 9 =	x 9 =	x =	
0 x 6 = 0	0 x 6 =	0 x =	x =	

Let's Practice!

Let's make sure you have all your facts down! Answer each multiplication problem.

$$\begin{array}{r} 0 \\ \times\ 6 \\ \hline \end{array}$$

$$\begin{array}{r} 0 \\ \times\ 10 \\ \hline \end{array}$$

$$\begin{array}{r} 0 \\ \times\ 2 \\ \hline \end{array}$$

$$\begin{array}{r} 0 \\ \times\ 4 \\ \hline \end{array}$$

$$\begin{array}{r} 0 \\ \times\ 7 \\ \hline \end{array}$$

$$\begin{array}{r} 0 \\ \times\ 8 \\ \hline \end{array}$$

$$\begin{array}{r} 0 \\ \times\ 3 \\ \hline \end{array}$$

$$\begin{array}{r} 0 \\ \times\ 1 \\ \hline \end{array}$$

$$\begin{array}{r} 0 \\ \times\ 5 \\ \hline \end{array}$$

$$\begin{array}{r} 0 \\ \times\ 9 \\ \hline \end{array}$$

$$\begin{array}{r} 0 \\ \times\ 0 \\ \hline \end{array}$$

I can multiply by 0:
TOP, BOTTOM AND MIXED

Name: _____ Date: _____

Goal: _____ problems in _____ seconds/minutes

1. 0 x 3	2. 0 x 8	3. 0 x 3	4. 0 x 2	5. 0 x 5	6. 0 x 4
7. 0 x 7	8. 0 x 2	9. 0 x 6	10. 0 x 0	11. 0 x 10	12. 0 x 1
13. 0 x 1	14. 0 x 6	15. 0 x 4	16. 0 x 9	17. 0 x 5	18. 0 x 7

19. 8 x 0	20. 10 x 0	21. 8 x 0	22. 1 x 0	23. 3 x 0	24. 0 x 0
25. 4 x 0	26. 1 x 0	27. 2 x 0	28. 10 x 0	29. 4 x 0	30. 6 x 0
31. 2 x 0	32. 9 x 0	33. 7 x 0	34. 9 x 0	35. 3 x 0	36. 5 x 0

37. 10 x 0	38. 3 x 0	39. 0 x 6	40. 0 x 2	41. 9 x 0	42. 0 x 5
43. 0 x 1	44. 0 x 5	45. 9 x 0	46. 4 x 0	47. 0 x 0	48. 0 x 2
49. 6 x 0	50. 0 x 3	51. 0 x 0	52. 0 x 7	53. 1 x 0	54. 8 x 0
55. 0 x 1	56. 3 x 0	57. 2 x 0	58. 0 x 8	59. 0 x 10	60. 10 x 0

Our New Fact Is ①

1 Facts
1 x 0 = 0
1 x 1 = 1
1 x 2 = 2
1 x 3 = 3
1 x 4 = 4
1 x 5 = 5
1 x 6 = 6
1 x 7 = 7
1 x 8 = 8
1 x 9 = 9
1 x 10 = 10

Draw straight lines from the facts to the correct answers!

1 x 3	2
1 x 10	4
1 x 4	0
1 x 1	8
1 x 6	3
1 x 9	7
1 x 5	9
1 x 0	10
1 x 8	5
1 x 2	1
1 x 7	6

Write Your 1 Facts

Trace it	Answer it	Fill in the blanks	Fill in the blanks	Write the fact
1 x 2 = 2	1 x 2 =	1 x =	x =	
1 x 7 = 7	1 x 7 =	x 7 =	x =	
1 x 3 = 3	1 x 3 =	1 x =	x =	
1 x 6 = 6	1 x 6 =	1 x =	x =	
1 x 8 = 8	1 x 8 =	1 x =	x =	
1 x 10 = 10	1 x 10 =	x 10 =	x =	
1 x 0 = 0	1 x 0 =	x 0 =	x =	
1 x 5 = 5	1 x 5 =	1 x =	x =	
1 x 9 = 9	1 x 9 =	1 x =	x =	
1 x 1 = 1	1 x 1 =	x 1 =	x =	
1 x 4 = 4	1 x 4 =	1 x =	x =	

Let's Practice!

Let's make sure you have all your facts down! Answer each multiplication problem.

$$\begin{array}{r} 1 \\ \times\ 4 \\ \hline \end{array}$$

$$\begin{array}{r} 1 \\ \times\ 7 \\ \hline \end{array}$$

$$\begin{array}{r} 1 \\ \times\ 2 \\ \hline \end{array}$$

$$\begin{array}{r} 1 \\ \times\ 9 \\ \hline \end{array}$$

$$\begin{array}{r} 1 \\ \times\ 5 \\ \hline \end{array}$$

$$\begin{array}{r} 1 \\ \times\ 8 \\ \hline \end{array}$$

$$\begin{array}{r} 1 \\ \times\ 0 \\ \hline \end{array}$$

$$\begin{array}{r} 1 \\ \times\ 3 \\ \hline \end{array}$$

$$\begin{array}{r} 1 \\ \times\ 6 \\ \hline \end{array}$$

$$\begin{array}{r} 1 \\ \times\ 10 \\ \hline \end{array}$$

$$\begin{array}{r} 1 \\ \times\ 1 \\ \hline \end{array}$$

I can multiply by 1: TOP or BOTTOM

1. 1×2	2. 1×10	3. 1×7	4. 1×4	5. 1×3	6. 1×9
7. 1×1	8. 1×0	9. 1×6	10. 1×6	11. 1×5	12. 1×7
13. 1×10	14. 1×8	15. 1×9	16. 1×2	17. 1×1	18. 1×3
19. 1×4	20. 1×9	21. 1×2	22. 1×1	23. 1×0	24. 1×8
25. 1×3	26. 1×4	27. 1×7	28. 1×2	29. 1×1	30. 1×5

---

31. 10×1	32. 5×1	33. 6×1	34. 2×1	35. 8×1	36. 6×1
37. 5×1	38. 6×1	39. 4×1	40. 0×1	41. 1×1	42. 3×1
43. 9×1	44. 10×1	45. 4×1	46. 6×1	47. 2×1	48. 5×1
49. 8×1	50. 9×1	51. 2×1	52. 2×1	53. 4×1	54. 7×1
55. 3×1	56. 3×1	57. 5×1	58. 4×1	59. 9×1	60. 0×1

I can multiply by 1s: Top AND Bottom

1. 1 x 5	2. 6 x 1	3. 9 x 1	4. 1 x 1	5. 0 x 1	6. 1 x 3
7. 6 x 1	8. 4 x 1	9. 1 x 8	10. 1 x 0	11. 2 x 1	12. 1 x 4
13. 3 x 1	14. 1 x 6	15. 1 x 9	16. 1 x 7	17. 8 x 1	18. 10 x 1
19. 1 x 8	20. 1 x 6	21. 3 x 1	22. 2 x 1	23. 1 x 1	24. 1 x 8
25. 1 x 0	26. 9 x 1	27. 1 x 3	28. 5 x 1	29. 1 x 6	30. 2 x 1
31. 0 x 1	32. 5 x 1	33. 1 x 3	34. 9 x 1	35. 7 x 1	36. 1 x 5
37. 4 x 1	38. 1 x 3	39. 1 x 8	40. 1 x 9	41. 0 x 1	42. 10 x 1
43. 1 x 2	44. 3 x 1	45. 6 x 1	46. 7 x 1	47. 1 x 8	48. 5 x 1
49. 4 x 1	50. 1 x 7	51. 8 x 1	52. 1 x 9	53. 1 x 1	54. 1 x 2
55. 1 x 3	56. 4 x 1	57. 8 x 1	58. 1 x 7	59. 5 x 1	60. 1 x 6

REVIEW: 0-1

1. 0 x 9	2. 1 x 2	3. 7 x 1	4. 1 x 4	5. 0 x 8	6. 9 x 0
7. 1 x 1	8. 1 x 7	9. 0 x 3	10. 10 x 0	11. 1 x 2	12. 8 x 1
13. 0 x 0	14. 1 x 3	15. 8 x 0	16. 0 x 4	17. 1 x 2	18. 1 x 8
19. 1 x 9	20. 2 x 0	21. 1 x 3	22. 1 x 9	23. 6 x 0	24. 2 x 0
25. 9 x 1	26. 1 x 1	27. 0 x 2	28. 9 x 0	29. 1 x 0	30. 0 x 10
31. 0 x 3	32. 0 x 5	33. 10 x 1	34. 1 x 2	35. 8 x 1	36. 0 x 4
37. 1 x 8	38. 1 x 2	39. 4 x 0	40. 6 x 0	41. 0 x 8	42. 1 x 6
43. 7 x 1	44. 3 x 1	45. 1 x 0	46. 1 x 0	47. 2 x 0	48. 5 x 0
49. 4 x 0	50. 7 x 1	51. 0 x 2	52. 1 x 8	53. 5 x 1	54. 0 x 8
55. 1 x 2	56. 0 x 9	57. 5 x 0	58. 1 x 3	59. 0 x 8	60. 6 x 1

REVIEW: 0-1

1. 1 × 4	2. 7 × 0	3. 0 × 3	4. 1 × 8	5. 10 × 1	6. 2 × 1
7. 0 × 4	8. 4 × 0	9. 2 × 0	10. 1 × 6	11. 3 × 1	12. 0 × 0
13. 1 × 10	14. 7 × 0	15. 3 × 1	16. 0 × 4	17. 1 × 1	18. 1 × 0
19. 0 × 4	20. 0 × 10	21. 8 × 1	22. 9 × 0	23. 0 × 3	24. 1 × 5
25. 0 × 9	26. 2 × 0	27. 1 × 8	28. 8 × 1	29. 0 × 2	30. 10 × 1
31. 8 × 1	32. 0 × 3	33. 7 × 1	34. 0 × 5	35. 6 × 0	36. 0 × 4
37. 5 × 1	38. 0 × 5	39. 8 × 1	40. 7 × 0	41. 0 × 9	42. 1 × 10
43. 7 × 0	44. 0 × 3	45. 3 × 1	46. 1 × 6	47. 1 × 3	48. 1 × 5
49. 5 × 0	50. 0 × 3	51. 0 × 4	52. 6 × 0	53. 1 × 3	54. 1 × 5
55. 7 × 1	56. 0 × 3	57. 1 × 7	58. 6 × 0	59. 10 × 0	60. 3 × 0

REVIEW: 0-1

1.	5	2.	0	3.	7	4.	9	5.	0	6.	4
x	1	x	2	x	1	x	0	x	2	x	1

7.	10	8.	8	9.	9	10.	1	11.	0	12.	0
x	1	x	0	x	1	x	4	x	4	x	2

13.	5	14.	7	15.	0	16.	6	17.	8	18.	1
x	0	x	1	x	4	x	1	x	0	x	3

19.	0	20.	0	21.	2	22.	0	23.	6	24.	4
x	3	x	10	x	1	x	3	x	0	x	0

25.	7	26.	0	27.	1	28.	7	29.	8	30.	0
x	0	x	3	x	6	x	1	x	0	x	3

31.	5	32.	0	33.	5	34.	1	35.	1	36.	0
x	0	x	8	x	1	x	1	x	7	x	3

37.	7	38.	10	39.	6	40.	3	41.	0	42.	1
x	1	x	0	x	0	x	1	x	5	x	2

43.	9	44.	3	45.	0	46.	8	47.	1	48.	8
x	0	x	1	x	4	x	1	x	6	x	1

49.	1	50.	9	51.	0	52.	8	53.	8	54.	0
x	0	x	0	x	7	x	1	x	0	x	3

55.	1	56.	5	57.	7	58.	0	59.	1	60.	8
x	10	x	0	x	1	x	9	x	2	x	1

Our New Fact Is **2**

2 Facts
2 x 0 = 0
2 x 1 = 2
2 x 2 = 4
2 x 3 = 6
2 x 4 = 8
2 x 5 = 10
2 x 6 = 12
2 x 7 = 14
2 x 8 = 16
2 x 9 = 18
2 x 10 = 20

Draw dashed lines from the facts to the correct answers!

2 x 3	10
2 x 9	8
2 x 2	20
2 x 7	6
2 x 6	0
2 x 1	16
2 x 10	12
2 x 8	4
2 x 0	14
2 x 5	18
2 x 4	2

Write Your 2 Facts

Trace it	Answer it	Fill in the blanks	Fill in the blanks	Write the fact
2 x 10 = 20	2 x 10 =	x 10 =	x =	
2 x 3 = 6	2 x 3 =	2 x =	x =	
2 x 6 = 12	2 x 6 =	2 x =	x =	
2 x 8 = 16	2 x 8 =	x 8 =	x =	
2 x 4 = 8	2 x 4 =	x 4 =	x =	
2 x 1 = 2	2 x 1 =	2 x =	x =	
2 x 9 = 18	2 x 9 =	2 x =	x =	
2 x 2 = 4	2 x 2 =	2 x =	x =	
2 x 0 = 0	2 x 0 =	x 0 =	x =	
2 x 5 = 10	2 x 5 =	2 x =	x =	
2 x 7 = 14	2 x 7 =	x 7 =	x =	

Let's Practice!

Let's make sure you have all your facts down! Answer each multiplication problem.

$$\begin{array}{r} 2 \\ \times\ 8 \\ \hline \end{array}$$

$$\begin{array}{r} 2 \\ \times\ 3 \\ \hline \end{array}$$

$$\begin{array}{r} 2 \\ \times\ 2 \\ \hline \end{array}$$

$$\begin{array}{r} 2 \\ \times\ 1 \\ \hline \end{array}$$

$$\begin{array}{r} 2 \\ \times\ 10 \\ \hline \end{array}$$

$$\begin{array}{r} 2 \\ \times\ 9 \\ \hline \end{array}$$

$$\begin{array}{r} 2 \\ \times\ 7 \\ \hline \end{array}$$

$$\begin{array}{r} 2 \\ \times\ 5 \\ \hline \end{array}$$

$$\begin{array}{r} 2 \\ \times\ 0 \\ \hline \end{array}$$

$$\begin{array}{r} 2 \\ \times\ 4 \\ \hline \end{array}$$

$$\begin{array}{r} 2 \\ \times\ 6 \\ \hline \end{array}$$

I can multiply by 2: TOP or BOTTOM

Name: _____ Date: _____

Goal: ____ problems in ____ seconds/minutes

1. 2 × 2	2. 2 × 1	3. 2 × 8	4. 2 × 10	5. 2 × 9	6. 2 × 5
7. 2 × 3	8. 2 × 7	9. 2 × 6	10. 2 × 8	11. 2 × 1	12. 2 × 0
13. 2 × 9	14. 2 × 8	15. 2 × 6	16. 2 × 5	17. 2 × 4	18. 2 × 8
19. 2 × 1	20. 2 × 2	21. 2 × 0	22. 2 × 10	23. 2 × 9	24. 2 × 3
25. 2 × 1	26. 2 × 2	27. 2 × 7	28. 2 × 7	29. 2 × 5	30. 2 × 3

---

31. 5 × 2	32. 4 × 2	33. 0 × 2	34. 10 × 2	35. 9 × 2	36. 6 × 2
37. 7 × 2	38. 4 × 2	39. 3 × 2	40. 2 × 2	41. 1 × 2	42. 0 × 2
43. 7 × 2	44. 9 × 2	45. 3 × 2	46. 4 × 2	47. 5 × 2	48. 5 × 2
49. 7 × 2	50. 2 × 2	51. 1 × 2	52. 8 × 2	53. 6 × 2	54. 9 × 2
55. 10 × 2	56. 2 × 2	57. 4 × 2	58. 7 × 2	59. 9 × 2	60. 3 × 2

I can multiply by 2s: Top AND Bottom

1. 4 × 2
2. 2 × 4
3. 2 × 3
4. 9 × 2
5. 10 × 2
6. 2 × 1

7. 2 × 0
8. 2 × 9
9. 8 × 2
10. 5 × 2
11. 2 × 4
12. 2 × 1

13. 3 × 2
14. 8 × 2
15. 2 × 10
16. 0 × 2
17. 2 × 1
18. 2 × 2

19. 8 × 2
20. 5 × 2
21. 3 × 2
22. 2 × 2
23. 2 × 7
24. 2 × 6

25. 2 × 6
26. 2 × 4
27. 2 × 9
28. 3 × 2
29. 2 × 2
30. 1 × 2

31. 0 × 2
32. 4 × 2
33. 2 × 5
34. 5 × 2
35. 2 × 7
36. 9 × 2

37. 2 × 5
38. 2 × 6
39. 8 × 2
40. 2 × 7
41. 2 × 2
42. 2 × 3

43. 2 × 10
44. 0 × 2
45. 2 × 2
46. 2 × 9
47. 2 × 4
48. 3 × 2

49. 6 × 2
50. 2 × 7
51. 2 × 8
52. 4 × 2
53. 9 × 2
54. 2 × 10

55. 5 × 2
56. 2 × 2
57. 2 × 1
58. 6 × 2
59. 2 × 4
60. 2 × 7

REVIEW: 0-2

Name: _____ Date: _____

Goal: _____ problems in _____ seconds/minutes

1. 2 x 2	2. 7 x 2	3. 0 x 1	4. 1 x 3	5. 4 x 2	6. 0 x 10
7. 9 x 2	8. 2 x 7	9. 7 x 1	10. 1 x 8	11. 0 x 2	12. 2 x 4
13. 6 x 0	14. 8 x 2	15. 0 x 5	16. 2 x 3	17. 3 x 1	18. 9 x 0
19. 0 x 3	20. 8 x 2	21. 2 x 7	22. 1 x 3	23. 0 x 2	24. 10 x 2
25. 6 x 2	26. 6 x 1	27. 7 x 2	28. 8 x 0	29. 0 x 2	30. 0 x 4
31. 4 x 1	32. 1 x 9	33. 2 x 7	34. 0 x 2	35. 2 x 9	36. 0 x 3
37. 9 x 1	38. 7 x 2	39. 0 x 5	40. 8 x 2	41. 1 x 6	42. 9 x 1
43. 2 x 7	44. 10 x 2	45. 0 x 8	46. 9 x 2	47. 5 x 2	48. 2 x 5
49. 3 x 2	50. 2 x 4	51. 0 x 6	52. 0 x 5	53. 1 x 2	54. 3 x 2
55. 2 x 10	56. 8 x 2	57. 6 x 1	58. 0 x 3	59. 3 x 1	60. 10 x 2

REVIEW: 0-2

1. $\begin{array}{r} 4 \\ \times\ 1 \\ \hline \end{array}$	2. $\begin{array}{r} 6 \\ \times\ 2 \\ \hline \end{array}$	3. $\begin{array}{r} 0 \\ \times\ 3 \\ \hline \end{array}$	4. $\begin{array}{r} 2 \\ \times\ 7 \\ \hline \end{array}$	5. $\begin{array}{r} 1 \\ \times\ 9 \\ \hline \end{array}$	6. $\begin{array}{r} 2 \\ \times\ 4 \\ \hline \end{array}$
7. $\begin{array}{r} 0 \\ \times\ 3 \\ \hline \end{array}$	8. $\begin{array}{r} 7 \\ \times\ 1 \\ \hline \end{array}$	9. $\begin{array}{r} 8 \\ \times\ 2 \\ \hline \end{array}$	10. $\begin{array}{r} 0 \\ \times\ 2 \\ \hline \end{array}$	11. $\begin{array}{r} 1 \\ \times\ 9 \\ \hline \end{array}$	12. $\begin{array}{r} 9 \\ \times\ 2 \\ \hline \end{array}$
13. $\begin{array}{r} 4 \\ \times\ 1 \\ \hline \end{array}$	14. $\begin{array}{r} 0 \\ \times\ 9 \\ \hline \end{array}$	15. $\begin{array}{r} 0 \\ \times\ 5 \\ \hline \end{array}$	16. $\begin{array}{r} 5 \\ \times\ 2 \\ \hline \end{array}$	17. $\begin{array}{r} 2 \\ \times\ 7 \\ \hline \end{array}$	18. $\begin{array}{r} 2 \\ \times\ 4 \\ \hline \end{array}$
19. $\begin{array}{r} 3 \\ \times\ 1 \\ \hline \end{array}$	20. $\begin{array}{r} 3 \\ \times\ 2 \\ \hline \end{array}$	21. $\begin{array}{r} 0 \\ \times\ 5 \\ \hline \end{array}$	22. $\begin{array}{r} 5 \\ \times\ 2 \\ \hline \end{array}$	23. $\begin{array}{r} 8 \\ \times\ 2 \\ \hline \end{array}$	24. $\begin{array}{r} 1 \\ \times\ 7 \\ \hline \end{array}$
25. $\begin{array}{r} 3 \\ \times\ 1 \\ \hline \end{array}$	26. $\begin{array}{r} 1 \\ \times\ 10 \\ \hline \end{array}$	27. $\begin{array}{r} 10 \\ \times\ 0 \\ \hline \end{array}$	28. $\begin{array}{r} 0 \\ \times\ 9 \\ \hline \end{array}$	29. $\begin{array}{r} 4 \\ \times\ 1 \\ \hline \end{array}$	30. $\begin{array}{r} 7 \\ \times\ 2 \\ \hline \end{array}$
31. $\begin{array}{r} 8 \\ \times\ 2 \\ \hline \end{array}$	32. $\begin{array}{r} 8 \\ \times\ 1 \\ \hline \end{array}$	33. $\begin{array}{r} 0 \\ \times\ 3 \\ \hline \end{array}$	34. $\begin{array}{r} 3 \\ \times\ 1 \\ \hline \end{array}$	35. $\begin{array}{r} 0 \\ \times\ 10 \\ \hline \end{array}$	36. $\begin{array}{r} 2 \\ \times\ 7 \\ \hline \end{array}$
37. $\begin{array}{r} 6 \\ \times\ 2 \\ \hline \end{array}$	38. $\begin{array}{r} 4 \\ \times\ 1 \\ \hline \end{array}$	39. $\begin{array}{r} 0 \\ \times\ 3 \\ \hline \end{array}$	40. $\begin{array}{r} 2 \\ \times\ 8 \\ \hline \end{array}$	41. $\begin{array}{r} 5 \\ \times\ 2 \\ \hline \end{array}$	42. $\begin{array}{r} 2 \\ \times\ 4 \\ \hline \end{array}$
43. $\begin{array}{r} 1 \\ \times\ 8 \\ \hline \end{array}$	44. $\begin{array}{r} 9 \\ \times\ 2 \\ \hline \end{array}$	45. $\begin{array}{r} 2 \\ \times\ 3 \\ \hline \end{array}$	46. $\begin{array}{r} 7 \\ \times\ 0 \\ \hline \end{array}$	47. $\begin{array}{r} 1 \\ \times\ 8 \\ \hline \end{array}$	48. $\begin{array}{r} 2 \\ \times\ 8 \\ \hline \end{array}$
49. $\begin{array}{r} 4 \\ \times\ 1 \\ \hline \end{array}$	50. $\begin{array}{r} 7 \\ \times\ 1 \\ \hline \end{array}$	51. $\begin{array}{r} 0 \\ \times\ 5 \\ \hline \end{array}$	52. $\begin{array}{r} 2 \\ \times\ 7 \\ \hline \end{array}$	53. $\begin{array}{r} 5 \\ \times\ 2 \\ \hline \end{array}$	54. $\begin{array}{r} 1 \\ \times\ 9 \\ \hline \end{array}$
55. $\begin{array}{r} 10 \\ \times\ 2 \\ \hline \end{array}$	56. $\begin{array}{r} 7 \\ \times\ 1 \\ \hline \end{array}$	57. $\begin{array}{r} 9 \\ \times\ 2 \\ \hline \end{array}$	58. $\begin{array}{r} 2 \\ \times\ 3 \\ \hline \end{array}$	59. $\begin{array}{r} 0 \\ \times\ 4 \\ \hline \end{array}$	60. $\begin{array}{r} 2 \\ \times\ 1 \\ \hline \end{array}$

REVIEW: 0-2

1. 7 x 1	2. 0 x 3	3. 2 x 2	4. 7 x 2	5. 0 x 3	6. 2 x 9
7. 8 x 0	8. 3 x 2	9. 0 x 0	10. 10 x 2	11. 1 x 1	12. 9 x 2
13. 1 x 8	14. 1 x 6	15. 0 x 3	16. 8 x 2	17. 5 x 0	18. 3 x 1
19. 0 x 3	20. 10 x 0	21. 2 x 2	22. 8 x 1	23. 1 x 6	24. 2 x 6
25. 8 x 1	26. 0 x 4	27. 0 x 7	28. 1 x 3	29. 2 x 6	30. 0 x 10
31. 6 x 2	32. 1 x 1	33. 7 x 2	34. 0 x 5	35. 2 x 4	36. 2 x 2
37. 1 x 2	38. 0 x 4	39. 3 x 1	40. 5 x 2	41. 5 x 0	42. 6 x 1
43. 1 x 8	44. 3 x 2	45. 1 x 10	46. 9 x 2	47. 9 x 1	48. 4 x 0
49. 6 x 2	50. 7 x 1	51. 0 x 4	52. 1 x 6	53. 10 x 2	54. 1 x 9
55. 7 x 2	56. 1 x 5	57. 2 x 4	58. 2 x 8	59. 1 x 3	60. 9 x 2

Our New Fact Is **3**

3 Facts
3 x 0 = 0
3 x 1 = 3
3 x 2 = 6
3 x 3 = 9
3 x 4 = 12
3 x 5 = 15
3 x 6 = 18
3 x 7 = 21
3 x 8 = 24
3 x 9 = 27
3 x 10 = 30

Draw squiggly lines from the facts to the correct answers!

Fact	Answer
3 x 5	15
3 x 1	21
3 x 2	24
3 x 7	12
3 x 10	18
3 x 9	27
3 x 3	3
3 x 8	9
3 x 6	30
3 x 4	6
3 x 0	0

Write Your 3 Facts

Trace it	Answer it	Fill in the blanks	Fill in the blanks	Write the fact
3 x 6 = 18	3 x 6 =	x 6 =	x =	
3 x 2 = 6	3 x 2 =	3 x =	x =	
3 x 7 = 14	3 x 7 =	3 x =	x =	
3 x 1 = 3	3 x 1 =	3 x =	x =	
3 x 5 = 15	3 x 5 =	x 5 =	x =	
3 x 8 = 24	3 x 8 =	x 8 =	x =	
3 x 4 = 12	3 x 4 =	3 x =	x =	
3 x 0 = 0	3 x 0 =	3 x =	x =	
3 x 9 = 27	3 x 9 =	x 9 =	x =	
3 x 3 = 9	3 x 3 =	3 x =	x =	
3 x 10 = 30	3 x 10 =	3 x =	x =	

Let's Practice!

Let's make sure you have all your facts down! Answer each multiplication problem.

$$3 \times 3$$

$$3 \times 7$$

$$3 \times 10$$

$$3 \times 2$$

$$3 \times 4$$

$$3 \times 5$$

$$3 \times 6$$

$$3 \times 9$$

$$3 \times 8$$

$$3 \times 2$$

$$3 \times 0$$

I can multiply by 3: TOP or BOTTOM

1. $\begin{array}{r} 3 \\ \times\ 5 \\ \hline \end{array}$	2. $\begin{array}{r} 3 \\ \times\ 2 \\ \hline \end{array}$	3. $\begin{array}{r} 3 \\ \times\ 3 \\ \hline \end{array}$	4. $\begin{array}{r} 3 \\ \times\ 1 \\ \hline \end{array}$	5. $\begin{array}{r} 3 \\ \times\ 0 \\ \hline \end{array}$	6. $\begin{array}{r} 3 \\ \times\ 10 \\ \hline \end{array}$
7. $\begin{array}{r} 3 \\ \times\ 6 \\ \hline \end{array}$	8. $\begin{array}{r} 3 \\ \times\ 9 \\ \hline \end{array}$	9. $\begin{array}{r} 3 \\ \times\ 7 \\ \hline \end{array}$	10. $\begin{array}{r} 3 \\ \times\ 3 \\ \hline \end{array}$	11. $\begin{array}{r} 3 \\ \times\ 2 \\ \hline \end{array}$	12. $\begin{array}{r} 3 \\ \times\ 7 \\ \hline \end{array}$
13. $\begin{array}{r} 3 \\ \times\ 9 \\ \hline \end{array}$	14. $\begin{array}{r} 3 \\ \times\ 5 \\ \hline \end{array}$	15. $\begin{array}{r} 3 \\ \times\ 4 \\ \hline \end{array}$	16. $\begin{array}{r} 3 \\ \times\ 1 \\ \hline \end{array}$	17. $\begin{array}{r} 3 \\ \times\ 0 \\ \hline \end{array}$	18. $\begin{array}{r} 3 \\ \times\ 10 \\ \hline \end{array}$
19. $\begin{array}{r} 3 \\ \times\ 9 \\ \hline \end{array}$	20. $\begin{array}{r} 3 \\ \times\ 8 \\ \hline \end{array}$	21. $\begin{array}{r} 3 \\ \times\ 3 \\ \hline \end{array}$	22. $\begin{array}{r} 3 \\ \times\ 4 \\ \hline \end{array}$	23. $\begin{array}{r} 3 \\ \times\ 2 \\ \hline \end{array}$	24. $\begin{array}{r} 3 \\ \times\ 8 \\ \hline \end{array}$
25. $\begin{array}{r} 3 \\ \times\ 9 \\ \hline \end{array}$	26. $\begin{array}{r} 3 \\ \times\ 7 \\ \hline \end{array}$	27. $\begin{array}{r} 3 \\ \times\ 4 \\ \hline \end{array}$	28. $\begin{array}{r} 3 \\ \times\ 6 \\ \hline \end{array}$	29. $\begin{array}{r} 3 \\ \times\ 8 \\ \hline \end{array}$	30. $\begin{array}{r} 3 \\ \times\ 2 \\ \hline \end{array}$

31. $\begin{array}{r} 1 \\ \times\ 3 \\ \hline \end{array}$	32. $\begin{array}{r} 0 \\ \times\ 3 \\ \hline \end{array}$	33. $\begin{array}{r} 10 \\ \times\ 3 \\ \hline \end{array}$	34. $\begin{array}{r} 4 \\ \times\ 3 \\ \hline \end{array}$	35. $\begin{array}{r} 8 \\ \times\ 3 \\ \hline \end{array}$	36. $\begin{array}{r} 9 \\ \times\ 3 \\ \hline \end{array}$
37. $\begin{array}{r} 3 \\ \times\ 3 \\ \hline \end{array}$	38. $\begin{array}{r} 4 \\ \times\ 3 \\ \hline \end{array}$	39. $\begin{array}{r} 0 \\ \times\ 3 \\ \hline \end{array}$	40. $\begin{array}{r} 10 \\ \times\ 3 \\ \hline \end{array}$	41. $\begin{array}{r} 8 \\ \times\ 3 \\ \hline \end{array}$	42. $\begin{array}{r} 3 \\ \times\ 3 \\ \hline \end{array}$
43. $\begin{array}{r} 9 \\ \times\ 3 \\ \hline \end{array}$	44. $\begin{array}{r} 8 \\ \times\ 3 \\ \hline \end{array}$	45. $\begin{array}{r} 3 \\ \times\ 3 \\ \hline \end{array}$	46. $\begin{array}{r} 2 \\ \times\ 3 \\ \hline \end{array}$	47. $\begin{array}{r} 6 \\ \times\ 3 \\ \hline \end{array}$	48. $\begin{array}{r} 7 \\ \times\ 3 \\ \hline \end{array}$
49. $\begin{array}{r} 2 \\ \times\ 3 \\ \hline \end{array}$	50. $\begin{array}{r} 2 \\ \times\ 3 \\ \hline \end{array}$	51. $\begin{array}{r} 6 \\ \times\ 3 \\ \hline \end{array}$	52. $\begin{array}{r} 9 \\ \times\ 3 \\ \hline \end{array}$	53. $\begin{array}{r} 5 \\ \times\ 3 \\ \hline \end{array}$	54. $\begin{array}{r} 1 \\ \times\ 3 \\ \hline \end{array}$
55. $\begin{array}{r} 10 \\ \times\ 3 \\ \hline \end{array}$	56. $\begin{array}{r} 9 \\ \times\ 3 \\ \hline \end{array}$	57. $\begin{array}{r} 2 \\ \times\ 3 \\ \hline \end{array}$	58. $\begin{array}{r} 4 \\ \times\ 3 \\ \hline \end{array}$	59. $\begin{array}{r} 3 \\ \times\ 3 \\ \hline \end{array}$	60. $\begin{array}{r} 7 \\ \times\ 3 \\ \hline \end{array}$

I can multiply by 3s: Top AND Bottom

Name: _____ Date: _____

Goal: ____ problems in ____ seconds/minutes

1. 9
  x 3

2. 3
  x 1

3. 1
  x 3

4. 3
  x 4

5. 6
  x 3

6. 3
  x 9

7. 3
  x 4

8. 3
  x 3

9. 8
  x 3

10. 3
  x 9

11. 7
  x 3

12. 4
  x 3

13. 3
  x 1

14. 0
  x 3

15. 10
  x 3

16. 9
  x 3

17. 3
  x 2

18. 4
  x 3

19. 5
  x 3

20. 7
  x 3

21. 3
  x 1

22. 8
  x 3

23. 3
  x 5

24. 3
  x 1

25. 3
  x 0

26. 3
  x 10

27. 8
  x 3

28. 3
  x 6

29. 9
  x 3

30. 3
  x 3

31. 1
  x 3

32. 3
  x 0

33. 4
  x 3

34. 3
  x 3

35. 3
  x 8

36. 7
  x 3

37. 3
  x 4

38. 5
  x 3

39. 3
  x 2

40. 3
  x 8

41. 3
  x 7

42. 4
  x 3

43. 1
  x 3

44. 0
  x 3

45. 3
  x 10

46. 5
  x 3

47. 3
  x 3

48. 7
  x 3

49. 3
  x 9

50. 3
  x 6

51. 2
  x 3

52. 3
  x 1

53. 0
  x 3

54. 3
  x 10

55. 7
  x 3

56. 3
  x 2

57. 3
  x 1

58. 0
  x 3

59. 3
  x 3

60. 3
  x 5

REVIEW: 0-3

1. $\begin{array}{r} 3 \\ \times\ 2 \\ \hline \end{array}$	2. $\begin{array}{r} 7 \\ \times\ 3 \\ \hline \end{array}$	3. $\begin{array}{r} 1 \\ \times\ 9 \\ \hline \end{array}$	4. $\begin{array}{r} 2 \\ \times\ 6 \\ \hline \end{array}$	5. $\begin{array}{r} 3 \\ \times\ 4 \\ \hline \end{array}$	6. $\begin{array}{r} 0 \\ \times\ 3 \\ \hline \end{array}$
7. $\begin{array}{r} 0 \\ \times\ 6 \\ \hline \end{array}$	8. $\begin{array}{r} 3 \\ \times\ 3 \\ \hline \end{array}$	9. $\begin{array}{r} 2 \\ \times\ 8 \\ \hline \end{array}$	10. $\begin{array}{r} 6 \\ \times\ 1 \\ \hline \end{array}$	11. $\begin{array}{r} 5 \\ \times\ 3 \\ \hline \end{array}$	12. $\begin{array}{r} 0 \\ \times\ 0 \\ \hline \end{array}$
13. $\begin{array}{r} 4 \\ \times\ 2 \\ \hline \end{array}$	14. $\begin{array}{r} 2 \\ \times\ 8 \\ \hline \end{array}$	15. $\begin{array}{r} 3 \\ \times\ 2 \\ \hline \end{array}$	16. $\begin{array}{r} 0 \\ \times\ 4 \\ \hline \end{array}$	17. $\begin{array}{r} 7 \\ \times\ 1 \\ \hline \end{array}$	18. $\begin{array}{r} 1 \\ \times\ 3 \\ \hline \end{array}$
19. $\begin{array}{r} 3 \\ \times\ 8 \\ \hline \end{array}$	20. $\begin{array}{r} 0 \\ \times\ 2 \\ \hline \end{array}$	21. $\begin{array}{r} 1 \\ \times\ 8 \\ \hline \end{array}$	22. $\begin{array}{r} 7 \\ \times\ 3 \\ \hline \end{array}$	23. $\begin{array}{r} 3 \\ \times\ 5 \\ \hline \end{array}$	24. $\begin{array}{r} 1 \\ \times\ 1 \\ \hline \end{array}$
25. $\begin{array}{r} 10 \\ \times\ 3 \\ \hline \end{array}$	26. $\begin{array}{r} 3 \\ \times\ 7 \\ \hline \end{array}$	27. $\begin{array}{r} 1 \\ \times\ 6 \\ \hline \end{array}$	28. $\begin{array}{r} 0 \\ \times\ 3 \\ \hline \end{array}$	29. $\begin{array}{r} 2 \\ \times\ 4 \\ \hline \end{array}$	30. $\begin{array}{r} 6 \\ \times\ 1 \\ \hline \end{array}$
31. $\begin{array}{r} 7 \\ \times\ 2 \\ \hline \end{array}$	32. $\begin{array}{r} 2 \\ \times\ 5 \\ \hline \end{array}$	33. $\begin{array}{r} 0 \\ \times\ 3 \\ \hline \end{array}$	34. $\begin{array}{r} 10 \\ \times\ 2 \\ \hline \end{array}$	35. $\begin{array}{r} 3 \\ \times\ 7 \\ \hline \end{array}$	36. $\begin{array}{r} 2 \\ \times\ 2 \\ \hline \end{array}$
37. $\begin{array}{r} 3 \\ \times\ 9 \\ \hline \end{array}$	38. $\begin{array}{r} 4 \\ \times\ 1 \\ \hline \end{array}$	39. $\begin{array}{r} 0 \\ \times\ 4 \\ \hline \end{array}$	40. $\begin{array}{r} 3 \\ \times\ 6 \\ \hline \end{array}$	41. $\begin{array}{r} 2 \\ \times\ 1 \\ \hline \end{array}$	42. $\begin{array}{r} 3 \\ \times\ 8 \\ \hline \end{array}$
43. $\begin{array}{r} 2 \\ \times\ 4 \\ \hline \end{array}$	44. $\begin{array}{r} 7 \\ \times\ 3 \\ \hline \end{array}$	45. $\begin{array}{r} 9 \\ \times\ 1 \\ \hline \end{array}$	46. $\begin{array}{r} 2 \\ \times\ 5 \\ \hline \end{array}$	47. $\begin{array}{r} 0 \\ \times\ 10 \\ \hline \end{array}$	48. $\begin{array}{r} 2 \\ \times\ 10 \\ \hline \end{array}$
49. $\begin{array}{r} 2 \\ \times\ 5 \\ \hline \end{array}$	50. $\begin{array}{r} 6 \\ \times\ 1 \\ \hline \end{array}$	51. $\begin{array}{r} 0 \\ \times\ 5 \\ \hline \end{array}$	52. $\begin{array}{r} 9 \\ \times\ 2 \\ \hline \end{array}$	53. $\begin{array}{r} 2 \\ \times\ 4 \\ \hline \end{array}$	54. $\begin{array}{r} 0 \\ \times\ 2 \\ \hline \end{array}$
55. $\begin{array}{r} 10 \\ \times\ 3 \\ \hline \end{array}$	56. $\begin{array}{r} 4 \\ \times\ 2 \\ \hline \end{array}$	57. $\begin{array}{r} 2 \\ \times\ 6 \\ \hline \end{array}$	58. $\begin{array}{r} 8 \\ \times\ 2 \\ \hline \end{array}$	59. $\begin{array}{r} 1 \\ \times\ 7 \\ \hline \end{array}$	60. $\begin{array}{r} 3 \\ \times\ 3 \\ \hline \end{array}$

REVIEW: 0-3

1. $\begin{array}{r} 6 \\ \times\ 2 \\ \hline \end{array}$	2. $\begin{array}{r} 7 \\ \times\ 1 \\ \hline \end{array}$	3. $\begin{array}{r} 0 \\ \times\ 3 \\ \hline \end{array}$	4. $\begin{array}{r} 2 \\ \times\ 2 \\ \hline \end{array}$	5. $\begin{array}{r} 1 \\ \times\ 0 \\ \hline \end{array}$	6. $\begin{array}{r} 10 \\ \times\ 1 \\ \hline \end{array}$
7. $\begin{array}{r} 3 \\ \times\ 7 \\ \hline \end{array}$	8. $\begin{array}{r} 1 \\ \times\ 5 \\ \hline \end{array}$	9. $\begin{array}{r} 8 \\ \times\ 1 \\ \hline \end{array}$	10. $\begin{array}{r} 9 \\ \times\ 0 \\ \hline \end{array}$	11. $\begin{array}{r} 2 \\ \times\ 4 \\ \hline \end{array}$	12. $\begin{array}{r} 0 \\ \times\ 3 \\ \hline \end{array}$
13. $\begin{array}{r} 1 \\ \times\ 4 \\ \hline \end{array}$	14. $\begin{array}{r} 3 \\ \times\ 6 \\ \hline \end{array}$	15. $\begin{array}{r} 8 \\ \times\ 2 \\ \hline \end{array}$	16. $\begin{array}{r} 9 \\ \times\ 3 \\ \hline \end{array}$	17. $\begin{array}{r} 1 \\ \times\ 10 \\ \hline \end{array}$	18. $\begin{array}{r} 0 \\ \times\ 4 \\ \hline \end{array}$
19. $\begin{array}{r} 5 \\ \times\ 0 \\ \hline \end{array}$	20. $\begin{array}{r} 10 \\ \times\ 3 \\ \hline \end{array}$	21. $\begin{array}{r} 6 \\ \times\ 2 \\ \hline \end{array}$	22. $\begin{array}{r} 3 \\ \times\ 4 \\ \hline \end{array}$	23. $\begin{array}{r} 3 \\ \times\ 8 \\ \hline \end{array}$	24. $\begin{array}{r} 2 \\ \times\ 0 \\ \hline \end{array}$
25. $\begin{array}{r} 7 \\ \times\ 1 \\ \hline \end{array}$	26. $\begin{array}{r} 5 \\ \times\ 0 \\ \hline \end{array}$	27. $\begin{array}{r} 2 \\ \times\ 7 \\ \hline \end{array}$	28. $\begin{array}{r} 8 \\ \times\ 2 \\ \hline \end{array}$	29. $\begin{array}{r} 1 \\ \times\ 0 \\ \hline \end{array}$	30. $\begin{array}{r} 2 \\ \times\ 2 \\ \hline \end{array}$
31. $\begin{array}{r} 4 \\ \times\ 3 \\ \hline \end{array}$	32. $\begin{array}{r} 3 \\ \times\ 3 \\ \hline \end{array}$	33. $\begin{array}{r} 2 \\ \times\ 8 \\ \hline \end{array}$	34. $\begin{array}{r} 4 \\ \times\ 0 \\ \hline \end{array}$	35. $\begin{array}{r} 0 \\ \times\ 6 \\ \hline \end{array}$	36. $\begin{array}{r} 10 \\ \times\ 2 \\ \hline \end{array}$
37. $\begin{array}{r} 2 \\ \times\ 9 \\ \hline \end{array}$	38. $\begin{array}{r} 4 \\ \times\ 1 \\ \hline \end{array}$	39. $\begin{array}{r} 0 \\ \times\ 7 \\ \hline \end{array}$	40. $\begin{array}{r} 10 \\ \times\ 3 \\ \hline \end{array}$	41. $\begin{array}{r} 9 \\ \times\ 3 \\ \hline \end{array}$	42. $\begin{array}{r} 3 \\ \times\ 9 \\ \hline \end{array}$
43. $\begin{array}{r} 2 \\ \times\ 5 \\ \hline \end{array}$	44. $\begin{array}{r} 5 \\ \times\ 3 \\ \hline \end{array}$	45. $\begin{array}{r} 0 \\ \times\ 9 \\ \hline \end{array}$	46. $\begin{array}{r} 6 \\ \times\ 0 \\ \hline \end{array}$	47. $\begin{array}{r} 1 \\ \times\ 3 \\ \hline \end{array}$	48. $\begin{array}{r} 3 \\ \times\ 8 \\ \hline \end{array}$
49. $\begin{array}{r} 0 \\ \times\ 4 \\ \hline \end{array}$	50. $\begin{array}{r} 2 \\ \times\ 8 \\ \hline \end{array}$	51. $\begin{array}{r} 1 \\ \times\ 8 \\ \hline \end{array}$	52. $\begin{array}{r} 9 \\ \times\ 3 \\ \hline \end{array}$	53. $\begin{array}{r} 2 \\ \times\ 3 \\ \hline \end{array}$	54. $\begin{array}{r} 5 \\ \times\ 1 \\ \hline \end{array}$
55. $\begin{array}{r} 10 \\ \times\ 0 \\ \hline \end{array}$	56. $\begin{array}{r} 0 \\ \times\ 6 \\ \hline \end{array}$	57. $\begin{array}{r} 3 \\ \times\ 5 \\ \hline \end{array}$	58. $\begin{array}{r} 2 \\ \times\ 8 \\ \hline \end{array}$	59. $\begin{array}{r} 1 \\ \times\ 4 \\ \hline \end{array}$	60. $\begin{array}{r} 3 \\ \times\ 2 \\ \hline \end{array}$

REVIEW: 0-3

Name: _____ Date: _____

Goal: _____ problems in _____ seconds/minutes

1. $\begin{array}{r} 0 \\ \times\ 0 \\ \hline \end{array}$	2. $\begin{array}{r} 8 \\ \times\ 3 \\ \hline \end{array}$	3. $\begin{array}{r} 2 \\ \times\ 5 \\ \hline \end{array}$	4. $\begin{array}{r} 0 \\ \times\ 2 \\ \hline \end{array}$	5. $\begin{array}{r} 10 \\ \times\ 2 \\ \hline \end{array}$	6. $\begin{array}{r} 2 \\ \times\ 7 \\ \hline \end{array}$
7. $\begin{array}{r} 2 \\ \times\ 8 \\ \hline \end{array}$	8. $\begin{array}{r} 8 \\ \times\ 2 \\ \hline \end{array}$	9. $\begin{array}{r} 1 \\ \times\ 2 \\ \hline \end{array}$	10. $\begin{array}{r} 4 \\ \times\ 0 \\ \hline \end{array}$	11. $\begin{array}{r} 0 \\ \times\ 5 \\ \hline \end{array}$	12. $\begin{array}{r} 3 \\ \times\ 1 \\ \hline \end{array}$
13. $\begin{array}{r} 4 \\ \times\ 3 \\ \hline \end{array}$	14. $\begin{array}{r} 6 \\ \times\ 2 \\ \hline \end{array}$	15. $\begin{array}{r} 2 \\ \times\ 7 \\ \hline \end{array}$	16. $\begin{array}{r} 3 \\ \times\ 4 \\ \hline \end{array}$	17. $\begin{array}{r} 8 \\ \times\ 3 \\ \hline \end{array}$	18. $\begin{array}{r} 2 \\ \times\ 9 \\ \hline \end{array}$
19. $\begin{array}{r} 1 \\ \times\ 6 \\ \hline \end{array}$	20. $\begin{array}{r} 3 \\ \times\ 7 \\ \hline \end{array}$	21. $\begin{array}{r} 8 \\ \times\ 3 \\ \hline \end{array}$	22. $\begin{array}{r} 10 \\ \times\ 0 \\ \hline \end{array}$	23. $\begin{array}{r} 3 \\ \times\ 0 \\ \hline \end{array}$	24. $\begin{array}{r} 3 \\ \times\ 2 \\ \hline \end{array}$
25. $\begin{array}{r} 2 \\ \times\ 4 \\ \hline \end{array}$	26. $\begin{array}{r} 8 \\ \times\ 2 \\ \hline \end{array}$	27. $\begin{array}{r} 9 \\ \times\ 3 \\ \hline \end{array}$	28. $\begin{array}{r} 1 \\ \times\ 5 \\ \hline \end{array}$	29. $\begin{array}{r} 4 \\ \times\ 1 \\ \hline \end{array}$	30. $\begin{array}{r} 3 \\ \times\ 7 \\ \hline \end{array}$
31. $\begin{array}{r} 10 \\ \times\ 1 \\ \hline \end{array}$	32. $\begin{array}{r} 1 \\ \times\ 1 \\ \hline \end{array}$	33. $\begin{array}{r} 5 \\ \times\ 2 \\ \hline \end{array}$	34. $\begin{array}{r} 7 \\ \times\ 3 \\ \hline \end{array}$	35. $\begin{array}{r} 3 \\ \times\ 9 \\ \hline \end{array}$	36. $\begin{array}{r} 2 \\ \times\ 5 \\ \hline \end{array}$
37. $\begin{array}{r} 2 \\ \times\ 2 \\ \hline \end{array}$	38. $\begin{array}{r} 4 \\ \times\ 1 \\ \hline \end{array}$	39. $\begin{array}{r} 1 \\ \times\ 7 \\ \hline \end{array}$	40. $\begin{array}{r} 9 \\ \times\ 2 \\ \hline \end{array}$	41. $\begin{array}{r} 0 \\ \times\ 7 \\ \hline \end{array}$	42. $\begin{array}{r} 3 \\ \times\ 0 \\ \hline \end{array}$
43. $\begin{array}{r} 2 \\ \times\ 4 \\ \hline \end{array}$	44. $\begin{array}{r} 6 \\ \times\ 2 \\ \hline \end{array}$	45. $\begin{array}{r} 7 \\ \times\ 3 \\ \hline \end{array}$	46. $\begin{array}{r} 0 \\ \times\ 8 \\ \hline \end{array}$	47. $\begin{array}{r} 2 \\ \times\ 0 \\ \hline \end{array}$	48. $\begin{array}{r} 1 \\ \times\ 6 \\ \hline \end{array}$
49. $\begin{array}{r} 9 \\ \times\ 2 \\ \hline \end{array}$	50. $\begin{array}{r} 10 \\ \times\ 3 \\ \hline \end{array}$	51. $\begin{array}{r} 6 \\ \times\ 3 \\ \hline \end{array}$	52. $\begin{array}{r} 2 \\ \times\ 9 \\ \hline \end{array}$	53. $\begin{array}{r} 3 \\ \times\ 5 \\ \hline \end{array}$	54. $\begin{array}{r} 2 \\ \times\ 3 \\ \hline \end{array}$
55. $\begin{array}{r} 1 \\ \times\ 6 \\ \hline \end{array}$	56. $\begin{array}{r} 6 \\ \times\ 3 \\ \hline \end{array}$	57. $\begin{array}{r} 2 \\ \times\ 8 \\ \hline \end{array}$	58. $\begin{array}{r} 9 \\ \times\ 0 \\ \hline \end{array}$	59. $\begin{array}{r} 10 \\ \times\ 2 \\ \hline \end{array}$	60. $\begin{array}{r} 6 \\ \times\ 2 \\ \hline \end{array}$

Our New Fact Is  4

4 Facts
4 x 0 = 0
4 x 1 = 4
4 x 2 = 8
4 x 3 = 12
4 x 4 = 16
4 x 5 = 20
4 x 6 = 24
4 x 7 = 28
4 x 8 = 32
4 x 9 = 36
4 x 10 = 40

Draw straight lines from the facts to the correct answers!

4 x 1	32
4 x 8	4
4 x 0	16
4 x 7	0
4 x 3	8
4 x 10	28
4 x 6	12
4 x 4	40
4 x 2	20
4 x 9	36
4 x 5	24

Write Your 4 Facts

Trace it	Answer it	Fill in the blanks	Fill in the blanks	Write the fact
4 x 2 = 8	4 x 2 =	x 2 =	x =	
4 x 7 = 28	4 x 7 =	4 x =	x =	
4 x 3 = 12	4 x 3 =	4 x =	x =	
4 x 6 = 24	4 x 6 =	x 6 =	x =	
4 x 8 = 32	4 x 8 =	x 8 =	x =	
4 x 10 = 40	4 x 10 =	4 x =	x =	
4 x 0 = 0	4 x 0 =	x 0 =	x =	
4 x 5 = 20	4 x 5 =	x 5 =	x =	
4 x 9 = 36	4 x 9 =	4 x =	x =	
4 x 1 = 4	4 x 1 =	x 1 =	x =	
4 x 4 = 16	4 x 4 =	x 4 =	x =	

Let's Practice!

Let's make sure you have all your facts down! Answer each multiplication problem.

$$\begin{array}{r} 4 \\ \times\ 5 \\ \hline \end{array}$$

$$\begin{array}{r} 4 \\ \times\ 1 \\ \hline \end{array}$$

$$\begin{array}{r} 4 \\ \times\ 8 \\ \hline \end{array}$$

$$\begin{array}{r} 4 \\ \times\ 6 \\ \hline \end{array}$$

$$\begin{array}{r} 4 \\ \times\ 2 \\ \hline \end{array}$$

$$\begin{array}{r} 4 \\ \times\ 4 \\ \hline \end{array}$$

$$\begin{array}{r} 4 \\ \times\ 9 \\ \hline \end{array}$$

$$\begin{array}{r} 4 \\ \times\ 0 \\ \hline \end{array}$$

$$\begin{array}{r} 4 \\ \times\ 3 \\ \hline \end{array}$$

$$\begin{array}{r} 4 \\ \times\ 10 \\ \hline \end{array}$$

$$\begin{array}{r} 4 \\ \times\ 7 \\ \hline \end{array}$$

I can multiply by 4: TOP or BOTTOM

1. 4 x 7	2. 4 x 2	3. 4 x 8	4. 4 x 0	5. 4 x 10	6. 4 x 2
7. 4 x 4	8. 4 x 5	9. 4 x 9	10. 4 x 7	11. 4 x 2	12. 4 x 4
13. 4 x 5	14. 4 x 0	15. 4 x 8	16. 4 x 1	17. 4 x 10	18. 4 x 3
19. 4 x 8	20. 4 x 9	21. 4 x 3	22. 4 x 5	23. 4 x 6	24. 4 x 8
25. 4 x 9	26. 4 x 3	27. 4 x 7	28. 4 x 5	29. 4 x 4	30. 4 x 1

31. 0 x 4	32. 8 x 4	33. 5 x 4	34. 3 x 4	35. 8 x 4	36. 9 x 4
37. 10 x 4	38. 2 x 4	39. 3 x 4	40. 9 x 4	41. 6 x 4	42. 4 x 4
43. 9 x 4	44. 8 x 4	45. 3 x 4	46. 5 x 4	47. 2 x 4	48. 4 x 4
49. 0 x 4	50. 10 x 4	51. 9 x 4	52. 3 x 4	53. 5 x 4	54. 8 x 4
55. 4 x 4	56. 6 x 4	57. 5 x 4	58. 9 x 4	59. 7 x 4	60. 3 x 4

I can multiply by 4s: Top AND Bottom

Name: _____ Date: _____

Goal: _____ problems in _____ seconds/minutes

1. 4 × 8	2. 4 × 4	3. 3 × 4
4. 4 × 9	5. 0 × 4	6. 4 × 10

7. 9 × 4	8. 6 × 4	9. 4 × 8
10. 4 × 2	11. 1 × 4	12. 4 × 0

13. 2 × 4	14. 4 × 3	15. 4 × 7
16. 4 × 9	17. 8 × 4	18. 4 × 4

19. 4 × 3	20. 4 × 10	21. 5 × 4
22. 8 × 4	23. 4 × 8	24. 4 × 3

25. 4 × 9	26. 7 × 4	27. 4 × 3
28. 6 × 4	29. 4 × 7	30. 4 × 4

31. 8 × 4	32. 1 × 4	33. 4 × 0
34. 10 × 4	35. 8 × 4	36. 4 × 3

37. 4 × 4	38. 4 × 6	39. 4 × 5
40. 4 × 8	41. 7 × 4	42. 6 × 4

43. 4 × 6	44. 1 × 4	45. 10 × 4
46. 0 × 4	47. 4 × 2	48. 4 × 4

49. 3 × 4	50. 4 × 8	51. 6 × 4
52. 4 × 5	53. 7 × 4	54. 4 × 3

55. 4 × 9	56. 8 × 4	57. 5 × 4
58. 4 × 3	59. 1 × 4	60. 4 × 2

REVIEW: 0-4

1. 4 x 2	2. 6 x 3	3. 2 x 8	4. 4 x 9	5. 10 x 3	6. 0 x 4
7. 3 x 4	8. 7 x 4	9. 1 x 9	10. 0 x 3	11. 6 x 1	12. 1 x 5
13. 0 x 4	14. 3 x 7	15. 8 x 4	16. 4 x 10	17. 2 x 7	18. 9 x 3
19. 5 x 4	20. 4 x 9	21. 3 x 6	22. 1 x 9	23. 4 x 2	24. 0 x 3
25. 8 x 0	26. 6 x 2	27. 1 x 5	28. 0 x 5	29. 1 x 4	30. 9 x 2
31. 7 x 4	32. 3 x 4	33. 8 x 4	34. 2 x 9	35. 10 x 3	36. 7 x 2
37. 0 x 6	38. 4 x 7	39. 1 x 8	40. 9 x 2	41. 4 x 5	42. 7 x 1
43. 0 x 0	44. 4 x 4	45. 9 x 4	46. 2 x 7	47. 3 x 9	48. 1 x 0
49. 8 x 2	50. 4 x 6	51. 0 x 9	52. 4 x 10	53. 3 x 6	54. 2 x 4
55. 7 x 1	56. 9 x 3	57. 1 x 8	58. 0 x 3	59. 2 x 6	60. 4 x 8

REVIEW: 0-4

Name: _____ Date: _____

Goal: _____ problems in _____ seconds/minutes

1. 7 x 2	2. 5 x 2	3. 2 x 9	4. 0 x 2	5. 1 x 7	6. 8 x 3
7. 4 x 4	8. 7 x 3	9. 4 x 9	10. 2 x 10	11. 3 x 7	12. 9 x 2
13. 6 x 4	14. 2 x 7	15. 3 x 9	16. 10 x 0	17. 3 x 3	18. 2 x 8
19. 5 x 3	20. 4 x 7	21. 2 x 8	22. 1 x 0	23. 2 x 6	24. 4 x 3
25. 9 x 3	26. 7 x 4	27. 2 x 2	28. 6 x 4	29. 0 x 9	30. 1 x 7
31. 1 x 1	32. 4 x 3	33. 9 x 4	34. 4 x 6	35. 2 x 0	36. 3 x 6
37. 10 x 3	38. 9 x 4	39. 2 x 6	40. 2 x 5	41. 0 x 5	42. 2 x 8
43. 7 x 2	44. 5 x 4	45. 1 x 8	46. 3 x 6	47. 10 x 4	48. 7 x 1
49. 0 x 3	50. 2 x 9	51. 4 x 0	52. 3 x 5	53. 8 x 4	54. 4 x 9
55. 9 x 4	56. 10 x 2	57. 8 x 1	58. 4 x 6	59. 7 x 2	60. 1 x 6

REVIEW: 0-4

Name: _____ Date: _____

Goal: _____ problems in _____ seconds/minutes

1. 6 × 3	2. 7 × 2	3. 0 × 8	4. 10 × 4	5. 2 × 0	6. 3 × 3
7. 0 × 7	8. 5 × 3	9. 2 × 8	10. 4 × 4	11. 3 × 3	12. 9 × 0
13. 1 × 7	14. 3 × 9	15. 4 × 7	16. 9 × 4	17. 2 × 5	18. 1 × 8
19. 0 × 5	20. 3 × 0	21. 2 × 7	22. 4 × 9	23. 0 × 2	24. 4 × 4
25. 4 × 0	26. 5 × 3	27. 4 × 8	28. 0 × 10	29. 7 × 3	30. 2 × 6
31. 5 × 4	32. 4 × 5	33. 8 × 2	34. 6 × 1	35. 3 × 10	36. 7 × 0
37. 6 × 3	38. 9 × 3	39. 0 × 6	40. 1 × 5	41. 4 × 8	42. 9 × 2
43. 3 × 6	44. 9 × 1	45. 0 × 7	46. 4 × 10	47. 7 × 3	48. 2 × 2
49. 2 × 9	50. 5 × 2	51. 1 × 7	52. 0 × 5	53. 3 × 1	54. 9 × 3
55. 10 × 2	56. 6 × 0	57. 8 × 2	58. 0 × 7	59. 3 × 7	60. 8 × 4

Our New Fact Is **5**

5 Facts
5 x 0 = 0
5 x 1 = 5
5 x 2 = 10
5 x 3 = 15
5 x 4 = 20
5 x 5 = 25
5 x 6 = 30
5 x 7 = 35
5 x 8 = 40
5 x 9 = 45
5 x 10 = 50

Draw dashed lines from the facts to the correct answers!

Fact	Answer
5 x 3	0
5 x 10	25
5 x 4	45
5 x 1	5
5 x 6	40
5 x 9	10
5 x 5	35
5 x 0	15
5 x 8	50
5 x 2	20
5 x 7	30

Write Your 5 Facts

Trace it	Answer it	Fill in the blanks	Fill in the blanks	Write the fact
5 x 3 = 15	5 x 3 =	x 3 =	x =	
5 x 4 = 20	5 x 4 =	5 x =	x =	
5 x 0 = 0	5 x 0 =	x 0 =	x =	
5 x 10 = 50	5 x 10 =	5 x =	x =	
5 x 5 = 25	5 x 5 =	5 x =	x =	
5 x 1 = 5	5 x 1 =	x 1 =	x =	
5 x 9 = 45	5 x 9 =	x 9 =	x =	
5 x 6 = 30	5 x 6 =	5 x =	x =	
5 x 2 = 10	5 x 2 =	x 2 =	x =	
5 x 8 = 40	5 x 8 =	5 x =	x =	
5 x 7 = 35	5 x 7 =	x 7 =	x =	

Let's Practice!

Let's make sure you have all your facts down! Answer each multiplication problem.

$$\begin{array}{r} 5 \\ \times\ 3 \\ \hline \end{array}$$

$$\begin{array}{r} 5 \\ \times\ 1 \\ \hline \end{array}$$

$$\begin{array}{r} 5 \\ \times\ 5 \\ \hline \end{array}$$

$$\begin{array}{r} 5 \\ \times\ 0 \\ \hline \end{array}$$

$$\begin{array}{r} 5 \\ \times\ 9 \\ \hline \end{array}$$

$$\begin{array}{r} 5 \\ \times\ 10 \\ \hline \end{array}$$

$$\begin{array}{r} 5 \\ \times\ 8 \\ \hline \end{array}$$

$$\begin{array}{r} 5 \\ \times\ 6 \\ \hline \end{array}$$

$$\begin{array}{r} 5 \\ \times\ 2 \\ \hline \end{array}$$

$$\begin{array}{r} 5 \\ \times\ 4 \\ \hline \end{array}$$

$$\begin{array}{r} 5 \\ \times\ 7 \\ \hline \end{array}$$

I can multiply by 5: TOP or BOTTOM

Name: _____ Date: _____

Goal: _____ problems in _____ seconds/minutes

1. $\begin{array}{r} 5 \\ \times\ 8 \\ \hline \end{array}$	2. $\begin{array}{r} 5 \\ \times\ 2 \\ \hline \end{array}$	3. $\begin{array}{r} 5 \\ \times\ 1 \\ \hline \end{array}$	4. $\begin{array}{r} 5 \\ \times\ 0 \\ \hline \end{array}$	5. $\begin{array}{r} 5 \\ \times\ 10 \\ \hline \end{array}$	6. $\begin{array}{r} 5 \\ \times\ 9 \\ \hline \end{array}$
7. $\begin{array}{r} 5 \\ \times\ 7 \\ \hline \end{array}$	8. $\begin{array}{r} 5 \\ \times\ 3 \\ \hline \end{array}$	9. $\begin{array}{r} 5 \\ \times\ 7 \\ \hline \end{array}$	10. $\begin{array}{r} 5 \\ \times\ 3 \\ \hline \end{array}$	11. $\begin{array}{r} 5 \\ \times\ 9 \\ \hline \end{array}$	12. $\begin{array}{r} 5 \\ \times\ 0 \\ \hline \end{array}$
13. $\begin{array}{r} 5 \\ \times\ 2 \\ \hline \end{array}$	14. $\begin{array}{r} 5 \\ \times\ 9 \\ \hline \end{array}$	15. $\begin{array}{r} 5 \\ \times\ 5 \\ \hline \end{array}$	16. $\begin{array}{r} 5 \\ \times\ 3 \\ \hline \end{array}$	17. $\begin{array}{r} 5 \\ \times\ 4 \\ \hline \end{array}$	18. $\begin{array}{r} 5 \\ \times\ 8 \\ \hline \end{array}$
19. $\begin{array}{r} 5 \\ \times\ 9 \\ \hline \end{array}$	20. $\begin{array}{r} 5 \\ \times\ 0 \\ \hline \end{array}$	21. $\begin{array}{r} 5 \\ \times\ 10 \\ \hline \end{array}$	22. $\begin{array}{r} 5 \\ \times\ 3 \\ \hline \end{array}$	23. $\begin{array}{r} 5 \\ \times\ 5 \\ \hline \end{array}$	24. $\begin{array}{r} 5 \\ \times\ 9 \\ \hline \end{array}$
25. $\begin{array}{r} 5 \\ \times\ 8 \\ \hline \end{array}$	26. $\begin{array}{r} 5 \\ \times\ 9 \\ \hline \end{array}$	27. $\begin{array}{r} 5 \\ \times\ 4 \\ \hline \end{array}$	28. $\begin{array}{r} 5 \\ \times\ 3 \\ \hline \end{array}$	29. $\begin{array}{r} 5 \\ \times\ 2 \\ \hline \end{array}$	30. $\begin{array}{r} 5 \\ \times\ 0 \\ \hline \end{array}$

31. $\begin{array}{r} 1 \\ \times\ 5 \\ \hline \end{array}$	32. $\begin{array}{r} 10 \\ \times\ 5 \\ \hline \end{array}$	33. $\begin{array}{r} 0 \\ \times\ 5 \\ \hline \end{array}$	34. $\begin{array}{r} 6 \\ \times\ 5 \\ \hline \end{array}$	35. $\begin{array}{r} 5 \\ \times\ 5 \\ \hline \end{array}$	36. $\begin{array}{r} 4 \\ \times\ 5 \\ \hline \end{array}$
37. $\begin{array}{r} 7 \\ \times\ 5 \\ \hline \end{array}$	38. $\begin{array}{r} 4 \\ \times\ 5 \\ \hline \end{array}$	39. $\begin{array}{r} 2 \\ \times\ 5 \\ \hline \end{array}$	40. $\begin{array}{r} 9 \\ \times\ 5 \\ \hline \end{array}$	41. $\begin{array}{r} 10 \\ \times\ 5 \\ \hline \end{array}$	42. $\begin{array}{r} 0 \\ \times\ 5 \\ \hline \end{array}$
43. $\begin{array}{r} 9 \\ \times\ 5 \\ \hline \end{array}$	44. $\begin{array}{r} 6 \\ \times\ 5 \\ \hline \end{array}$	45. $\begin{array}{r} 3 \\ \times\ 5 \\ \hline \end{array}$	46. $\begin{array}{r} 5 \\ \times\ 5 \\ \hline \end{array}$	47. $\begin{array}{r} 7 \\ \times\ 5 \\ \hline \end{array}$	48. $\begin{array}{r} 6 \\ \times\ 5 \\ \hline \end{array}$
49. $\begin{array}{r} 2 \\ \times\ 5 \\ \hline \end{array}$	50. $\begin{array}{r} 3 \\ \times\ 5 \\ \hline \end{array}$	51. $\begin{array}{r} 8 \\ \times\ 5 \\ \hline \end{array}$	52. $\begin{array}{r} 7 \\ \times\ 5 \\ \hline \end{array}$	53. $\begin{array}{r} 9 \\ \times\ 5 \\ \hline \end{array}$	54. $\begin{array}{r} 0 \\ \times\ 5 \\ \hline \end{array}$
55. $\begin{array}{r} 10 \\ \times\ 5 \\ \hline \end{array}$	56. $\begin{array}{r} 8 \\ \times\ 5 \\ \hline \end{array}$	57. $\begin{array}{r} 9 \\ \times\ 5 \\ \hline \end{array}$	58. $\begin{array}{r} 3 \\ \times\ 5 \\ \hline \end{array}$	59. $\begin{array}{r} 5 \\ \times\ 5 \\ \hline \end{array}$	60. $\begin{array}{r} 7 \\ \times\ 5 \\ \hline \end{array}$

I can multiply by 5s: Top AND Bottom

Name: _____ Date: _____

Goal: _____ problems in _____ seconds/minutes

1. 5 × 5

2. 5 × 0

3. 5 × 6

4. 5 × 5

5. 8 × 5

6. 5 × 8

7. 5 × 9

8. 5 × 1

9. 2 × 5

10. 8 × 5

11. 5 × 7

12. 5 × 9

13. 10 × 5

14. 8 × 5

15. 5 × 9

16. 3 × 5

17. 5 × 7

18. 9 × 5

19. 1 × 5

20. 0 × 5

21. 3 × 5

22. 5 × 2

23. 5 × 5

24. 5 × 7

25. 5 × 6

26. 5 × 1

27. 5 × 0

28. 10 × 5

29. 5 × 8

30. 3 × 5

31. 2 × 5

32. 4 × 5

33. 5 × 4

34. 7 × 5

35. 5 × 6

36. 8 × 5

37. 5 × 2

38. 5 × 7

39. 0 × 5

40. 5 × 2

41. 7 × 5

42. 5 × 9

43. 5 × 8

44. 6 × 5

45. 3 × 5

46. 5 × 9

47. 5 × 4

48. 3 × 5

49. 1 × 5

50. 5 × 7

51. 5 × 9

52. 3 × 5

53. 2 × 5

54. 5 × 7

55. 5 × 5

56. 6 × 5

57. 5 × 9

58. 10 × 5

59. 5 × 3

60. 5 × 9

REVIEW: 0-5

1. $\begin{array}{r} 1 \\ \times\ 7 \\ \hline \end{array}$	2. $\begin{array}{r} 5 \\ \times\ 3 \\ \hline \end{array}$	3. $\begin{array}{r} 8 \\ \times\ 3 \\ \hline \end{array}$	4. $\begin{array}{r} 2 \\ \times\ 7 \\ \hline \end{array}$	5. $\begin{array}{r} 6 \\ \times\ 5 \\ \hline \end{array}$	6. $\begin{array}{r} 0 \\ \times\ 8 \\ \hline \end{array}$
7. $\begin{array}{r} 2 \\ \times\ 3 \\ \hline \end{array}$	8. $\begin{array}{r} 5 \\ \times\ 9 \\ \hline \end{array}$	9. $\begin{array}{r} 7 \\ \times\ 1 \\ \hline \end{array}$	10. $\begin{array}{r} 0 \\ \times\ 4 \\ \hline \end{array}$	11. $\begin{array}{r} 4 \\ \times\ 8 \\ \hline \end{array}$	12. $\begin{array}{r} 5 \\ \times\ 5 \\ \hline \end{array}$
13. $\begin{array}{r} 3 \\ \times\ 4 \\ \hline \end{array}$	14. $\begin{array}{r} 4 \\ \times\ 7 \\ \hline \end{array}$	15. $\begin{array}{r} 0 \\ \times\ 6 \\ \hline \end{array}$	16. $\begin{array}{r} 6 \\ \times\ 3 \\ \hline \end{array}$	17. $\begin{array}{r} 9 \\ \times\ 3 \\ \hline \end{array}$	18. $\begin{array}{r} 10 \\ \times\ 5 \\ \hline \end{array}$
19. $\begin{array}{r} 10 \\ \times\ 3 \\ \hline \end{array}$	20. $\begin{array}{r} 0 \\ \times\ 7 \\ \hline \end{array}$	21. $\begin{array}{r} 3 \\ \times\ 9 \\ \hline \end{array}$	22. $\begin{array}{r} 5 \\ \times\ 2 \\ \hline \end{array}$	23. $\begin{array}{r} 2 \\ \times\ 8 \\ \hline \end{array}$	24. $\begin{array}{r} 0 \\ \times\ 3 \\ \hline \end{array}$
25. $\begin{array}{r} 9 \\ \times\ 0 \\ \hline \end{array}$	26. $\begin{array}{r} 4 \\ \times\ 4 \\ \hline \end{array}$	27. $\begin{array}{r} 6 \\ \times\ 5 \\ \hline \end{array}$	28. $\begin{array}{r} 5 \\ \times\ 9 \\ \hline \end{array}$	29. $\begin{array}{r} 2 \\ \times\ 6 \\ \hline \end{array}$	30. $\begin{array}{r} 1 \\ \times\ 5 \\ \hline \end{array}$
31. $\begin{array}{r} 4 \\ \times\ 3 \\ \hline \end{array}$	32. $\begin{array}{r} 8 \\ \times\ 5 \\ \hline \end{array}$	33. $\begin{array}{r} 0 \\ \times\ 2 \\ \hline \end{array}$	34. $\begin{array}{r} 1 \\ \times\ 10 \\ \hline \end{array}$	35. $\begin{array}{r} 3 \\ \times\ 8 \\ \hline \end{array}$	36. $\begin{array}{r} 5 \\ \times\ 8 \\ \hline \end{array}$
37. $\begin{array}{r} 4 \\ \times\ 9 \\ \hline \end{array}$	38. $\begin{array}{r} 7 \\ \times\ 2 \\ \hline \end{array}$	39. $\begin{array}{r} 9 \\ \times\ 3 \\ \hline \end{array}$	40. $\begin{array}{r} 5 \\ \times\ 10 \\ \hline \end{array}$	41. $\begin{array}{r} 4 \\ \times\ 2 \\ \hline \end{array}$	42. $\begin{array}{r} 8 \\ \times\ 3 \\ \hline \end{array}$
43. $\begin{array}{r} 4 \\ \times\ 0 \\ \hline \end{array}$	44. $\begin{array}{r} 10 \\ \times\ 2 \\ \hline \end{array}$	45. $\begin{array}{r} 8 \\ \times\ 5 \\ \hline \end{array}$	46. $\begin{array}{r} 3 \\ \times\ 7 \\ \hline \end{array}$	47. $\begin{array}{r} 3 \\ \times\ 9 \\ \hline \end{array}$	48. $\begin{array}{r} 0 \\ \times\ 1 \\ \hline \end{array}$
49. $\begin{array}{r} 3 \\ \times\ 7 \\ \hline \end{array}$	50. $\begin{array}{r} 9 \\ \times\ 2 \\ \hline \end{array}$	51. $\begin{array}{r} 6 \\ \times\ 5 \\ \hline \end{array}$	52. $\begin{array}{r} 4 \\ \times\ 4 \\ \hline \end{array}$	53. $\begin{array}{r} 8 \\ \times\ 0 \\ \hline \end{array}$	54. $\begin{array}{r} 2 \\ \times\ 7 \\ \hline \end{array}$
55. $\begin{array}{r} 3 \\ \times\ 10 \\ \hline \end{array}$	56. $\begin{array}{r} 0 \\ \times\ 2 \\ \hline \end{array}$	57. $\begin{array}{r} 8 \\ \times\ 5 \\ \hline \end{array}$	58. $\begin{array}{r} 5 \\ \times\ 8 \\ \hline \end{array}$	59. $\begin{array}{r} 3 \\ \times\ 9 \\ \hline \end{array}$	60. $\begin{array}{r} 2 \\ \times\ 2 \\ \hline \end{array}$

REVIEW: 0-5

Name: _____ Date: _____

Goal: _____ problems in _____ seconds/minutes

1. 3×4	2. 7×2	3. 0×7	4. 4×6	5. 9×1	6. 8×5
7. 7×4	8. 10×5	9. 6×3	10. 3×6	11. 8×4	12. 5×9
13. 2×1	14. 0×8	15. 4×5	16. 3×7	17. 8×2	18. 1×9
19. 3×10	20. 5×6	21. 9×4	22. 5×5	23. 6×2	24. 2×8
25. 0×6	26. 4×0	27. 3×3	28. 8×4	29. 1×9	30. 8×5
31. 3×3	32. 2×3	33. 6×4	34. 1×3	35. 10×3	36. 0×0
37. 5×4	38. 7×5	39. 3×9	40. 2×7	41. 0×1	42. 7×3
43. 0×7	44. 5×10	45. 6×4	46. 2×9	47. 3×8	48. 7×4
49. 3×3	50. 4×2	51. 2×5	52. 7×0	53. 1×0	54. 2×8
55. 10×4	56. 8×5	57. 3×9	58. 7×2	59. 5×4	60. 2×3

REVIEW: 0-5

Name: _____ Date: _____

Goal: _____ problems in _____ seconds/minutes

1. 8×5	2. 4×3	3. 4×4	4. 8×2	5. 10×3	6. 7×0
7. 6×5	8. 4×8	9. 9×2	10. 1×0	11. 2×10	12. 7×2
13. 2×2	14. 5×2	15. 5×7	16. 2×9	17. 3×7	18. 5×5
19. 4×5	20. 4×10	21. 0×7	22. 6×4	23. 4×8	24. 7×3
25. 2×1	26. 4×1	27. 1×1	28. 0×8	29. 7×5	30. 4×8
31. 0×6	32. 4×6	33. 2×8	34. 7×1	35. 4×6	36. 6×2
37. 9×5	38. 2×0	39. 3×5	40. 6×3	41. 9×1	42. 4×8
43. 6×5	44. 4×8	45. 2×9	46. 0×1	47. 10×5	48. 4×3
49. 3×4	50. 4×7	51. 2×9	52. 8×5	53. 4×7	54. 10×4
55. 1×5	56. 6×5	57. 4×5	58. 7×3	59. 2×4	60. 5×9

Our New Fact Is 6

6 Facts
6 x 0 = 0
6 x 1 = 6
6 x 2 = 12
6 x 3 = 18
6 x 4 = 24
6 x 5 = 32
6 x 6 = 36
6 x 7 = 42
6 x 8 = 48
6 x 9 = 54
6 x 10 = 60

Draw squiggly lines from the facts to the correct answers!

6 x 5	12
6 x 6	54
6 x 1	48
6 x 3	18
6 x 8	0
6 x 4	24
6 x 0	42
6 x 9	60
6 x 2	32
6 x 7	36
6 x 10	6

Write Your 6 Facts

Trace it	Answer it	Fill in the blanks	Fill in the blanks	Write the fact
6 x 7 = 42	6 x 7 =	6 x __ =	__ x __ =	
6 x 3 = 18	6 x 3 =	__ x 3 =	__ x __ =	
6 x 6 = 36	6 x 6 =	6 x __ =	__ x __ =	
6 x 2 = 12	6 x 2 =	__ x 2 =	__ x __ =	
6 x 4 = 24	6 x 4 =	__ x 4 =	__ x __ =	
6 x 8 = 18	6 x 8 =	6 x __ =	__ x __ =	
6 x 1 = 6	6 x 1 =	__ x 1 =	__ x __ =	
6 x 9 = 54	6 x 9 =	6 x __ =	__ x __ =	
6 x 5 = 30	6 x 5 =	6 x __ =	__ x __ =	
6 x 10 = 60	6 x 10 =	__ x 10 =	__ x __ =	
6 x 0 = 0	6 x 0 =	__ x 0 =	__ x __ =	

Let's Practice!

Let's make sure you have all your facts down! Answer each multiplication problem.

$$6 \times 6$$

$$6 \times 7$$

$$6 \times 8$$

$$6 \times 5$$

$$6 \times 9$$

$$6 \times 0$$

$$6 \times 4$$

$$6 \times 1$$

$$6 \times 3$$

$$6 \times 2$$

$$6 \times 10$$

I can multiply by 6: TOP or BOTTOM

Name: _____ Date: _____

Goal: _____ problems in _____ seconds/minutes

1. 6 × 5	2. 6 × 7	3. 6 × 0	4. 6 × 10	5. 6 × 3	6. 6 × 2
7. 6 × 8	8. 6 × 6	9. 6 × 4	10. 6 × 6	11. 6 × 5	12. 6 × 8
13. 6 × 9	14. 6 × 10	15. 6 × 0	16. 6 × 1	17. 6 × 2	18. 6 × 5
19. 6 × 9	20. 6 × 8	21. 6 × 5	22. 6 × 3	23. 6 × 7	24. 6 × 8
25. 6 × 10	26. 6 × 2	27. 6 × 9	28. 6 × 1	29. 6 × 7	30. 6 × 5

---

31. 3 × 6	32. 8 × 6	33. 7 × 6	34. 5 × 6	35. 4 × 6	36. 10 × 6
37. 7 × 6	38. 3 × 6	39. 2 × 6	40. 3 × 6	41. 8 × 6	42. 7 × 6
43. 10 × 6	44. 0 × 6	45. 2 × 6	46. 1 × 6	47. 8 × 6	48. 9 × 6
49. 6 × 6	50. 3 × 6	51. 6 × 6	52. 9 × 6	53. 7 × 6	54. 1 × 6
55. 10 × 6	56. 0 × 6	57. 4 × 6	58. 9 × 6	59. 2 × 6	60. 4 × 6

I can multiply by 6s: Top AND Bottom

1. 10 × 6	2. 6 × 0	3. 1 × 6	4. 6 × 8	5. 9 × 6	6. 6 × 5
7. 6 × 2	8. 6 × 1	9. 7 × 6	10. 6 × 9	11. 10 × 6	12. 2 × 6
13. 6 × 9	14. 7 × 6	15. 4 × 6	16. 6 × 6	17. 6 × 1	18. 5 × 6
19. 9 × 6	20. 7 × 6	21. 6 × 3	22. 5 × 6	23. 6 × 8	24. 6 × 2
25. 6 × 8	26. 6 × 0	27. 10 × 6	28. 6 × 8	29. 9 × 6	30. 6 × 6
31. 2 × 6	32. 6 × 8	33. 2 × 6	34. 4 × 6	35. 6 × 7	36. 9 × 6
37. 6 × 10	38. 8 × 6	39. 6 × 3	40. 6 × 5	41. 6 × 3	42. 8 × 6
43. 9 × 6	44. 4 × 6	45. 6 × 3	46. 7 × 6	47. 9 × 6	48. 1 × 6
49. 6 × 0	50. 6 × 10	51. 6 × 6	52. 6 × 4	53. 7 × 6	54. 6 × 2
55. 8 × 6	56. 6 × 7	57. 6 × 5	58. 9 × 6	59. 6 × 6	60. 6 × 1

REVIEW: 0-6

Name: _____ Date: _____

Goal: _____ problems in _____ seconds/minutes

1. 7 × 6	2. 3 × 6	3. 8 × 1	4. 9 × 0	5. 2 × 6	6. 6 × 4
7. 5 × 3	8. 8 × 2	9. 10 × 6	10. 2 × 3	11. 7 × 3	12. 5 × 9
13. 2 × 10	14. 3 × 0	15. 5 × 8	16. 6 × 6	17. 9 × 6	18. 3 × 1
19. 8 × 3	20. 4 × 6	21. 0 × 1	22. 5 × 8	23. 3 × 0	24. 6 × 10
25. 8 × 5	26. 3 × 7	27. 2 × 9	28. 8 × 4	29. 6 × 5	30. 0 × 9
31. 1 × 6	32. 8 × 0	33. 5 × 7	34. 6 × 7	35. 9 × 2	36. 8 × 3
37. 4 × 4	38. 3 × 7	39. 3 × 9	40. 2 × 0	41. 9 × 4	42. 5 × 5
43. 2 × 1	44. 1 × 7	45. 3 × 9	46. 0 × 4	47. 5 × 6	48. 8 × 3
49. 9 × 2	50. 4 × 7	51. 6 × 7	52. 9 × 3	53. 6 × 8	54. 2 × 2
55. 5 × 2	56. 4 × 5	57. 6 × 4	58. 4 × 6	59. 9 × 6	60. 3 × 3

REVIEW: 0-6

1. $\begin{array}{r} 5 \\ \times\ 5 \\ \hline \end{array}$	2. $\begin{array}{r} 3 \\ \times\ 6 \\ \hline \end{array}$	3. $\begin{array}{r} 9 \\ \times\ 2 \\ \hline \end{array}$	4. $\begin{array}{r} 0 \\ \times\ 4 \\ \hline \end{array}$	5. $\begin{array}{r} 5 \\ \times\ 9 \\ \hline \end{array}$	6. $\begin{array}{r} 8 \\ \times\ 4 \\ \hline \end{array}$
7. $\begin{array}{r} 1 \\ \times\ 3 \\ \hline \end{array}$	8. $\begin{array}{r} 3 \\ \times\ 5 \\ \hline \end{array}$	9. $\begin{array}{r} 2 \\ \times\ 2 \\ \hline \end{array}$	10. $\begin{array}{r} 7 \\ \times\ 0 \\ \hline \end{array}$	11. $\begin{array}{r} 9 \\ \times\ 6 \\ \hline \end{array}$	12. $\begin{array}{r} 8 \\ \times\ 5 \\ \hline \end{array}$
13. $\begin{array}{r} 0 \\ \times\ 10 \\ \hline \end{array}$	14. $\begin{array}{r} 7 \\ \times\ 6 \\ \hline \end{array}$	15. $\begin{array}{r} 5 \\ \times\ 9 \\ \hline \end{array}$	16. $\begin{array}{r} 3 \\ \times\ 6 \\ \hline \end{array}$	17. $\begin{array}{r} 6 \\ \times\ 3 \\ \hline \end{array}$	18. $\begin{array}{r} 2 \\ \times\ 6 \\ \hline \end{array}$
19. $\begin{array}{r} 9 \\ \times\ 4 \\ \hline \end{array}$	20. $\begin{array}{r} 4 \\ \times\ 4 \\ \hline \end{array}$	21. $\begin{array}{r} 6 \\ \times\ 6 \\ \hline \end{array}$	22. $\begin{array}{r} 2 \\ \times\ 9 \\ \hline \end{array}$	23. $\begin{array}{r} 0 \\ \times\ 8 \\ \hline \end{array}$	24. $\begin{array}{r} 5 \\ \times\ 3 \\ \hline \end{array}$
25. $\begin{array}{r} 1 \\ \times\ 7 \\ \hline \end{array}$	26. $\begin{array}{r} 7 \\ \times\ 5 \\ \hline \end{array}$	27. $\begin{array}{r} 6 \\ \times\ 4 \\ \hline \end{array}$	28. $\begin{array}{r} 9 \\ \times\ 3 \\ \hline \end{array}$	29. $\begin{array}{r} 2 \\ \times\ 7 \\ \hline \end{array}$	30. $\begin{array}{r} 6 \\ \times\ 3 \\ \hline \end{array}$
31. $\begin{array}{r} 7 \\ \times\ 0 \\ \hline \end{array}$	32. $\begin{array}{r} 3 \\ \times\ 10 \\ \hline \end{array}$	33. $\begin{array}{r} 9 \\ \times\ 0 \\ \hline \end{array}$	34. $\begin{array}{r} 2 \\ \times\ 6 \\ \hline \end{array}$	35. $\begin{array}{r} 7 \\ \times\ 3 \\ \hline \end{array}$	36. $\begin{array}{r} 9 \\ \times\ 6 \\ \hline \end{array}$
37. $\begin{array}{r} 6 \\ \times\ 7 \\ \hline \end{array}$	38. $\begin{array}{r} 9 \\ \times\ 3 \\ \hline \end{array}$	39. $\begin{array}{r} 7 \\ \times\ 2 \\ \hline \end{array}$	40. $\begin{array}{r} 6 \\ \times\ 5 \\ \hline \end{array}$	41. $\begin{array}{r} 5 \\ \times\ 5 \\ \hline \end{array}$	42. $\begin{array}{r} 9 \\ \times\ 2 \\ \hline \end{array}$
43. $\begin{array}{r} 2 \\ \times\ 6 \\ \hline \end{array}$	44. $\begin{array}{r} 5 \\ \times\ 7 \\ \hline \end{array}$	45. $\begin{array}{r} 3 \\ \times\ 0 \\ \hline \end{array}$	46. $\begin{array}{r} 10 \\ \times\ 6 \\ \hline \end{array}$	47. $\begin{array}{r} 4 \\ \times\ 5 \\ \hline \end{array}$	48. $\begin{array}{r} 3 \\ \times\ 8 \\ \hline \end{array}$
49. $\begin{array}{r} 2 \\ \times\ 2 \\ \hline \end{array}$	50. $\begin{array}{r} 6 \\ \times\ 2 \\ \hline \end{array}$	51. $\begin{array}{r} 8 \\ \times\ 3 \\ \hline \end{array}$	52. $\begin{array}{r} 9 \\ \times\ 5 \\ \hline \end{array}$	53. $\begin{array}{r} 5 \\ \times\ 8 \\ \hline \end{array}$	54. $\begin{array}{r} 3 \\ \times\ 7 \\ \hline \end{array}$
55. $\begin{array}{r} 2 \\ \times\ 8 \\ \hline \end{array}$	56. $\begin{array}{r} 10 \\ \times\ 5 \\ \hline \end{array}$	57. $\begin{array}{r} 4 \\ \times\ 6 \\ \hline \end{array}$	58. $\begin{array}{r} 7 \\ \times\ 2 \\ \hline \end{array}$	59. $\begin{array}{r} 4 \\ \times\ 9 \\ \hline \end{array}$	60. $\begin{array}{r} 6 \\ \times\ 6 \\ \hline \end{array}$

REVIEW: 0-6

1. 9 × 4	2. 0 × 5	3. 2 × 9	4. 7 × 3	5. 4 × 6	6. 2 × 2
7. 7 × 6	8. 4 × 8	9. 9 × 2	10. 1 × 0	11. 1 × 1	12. 8 × 3
13. 2 × 8	14. 3 × 3	15. 5 × 8	16. 6 × 7	17. 5 × 5	18. 9 × 3
19. 3 × 2	20. 2 × 2	21. 7 × 4	22. 1 × 5	23. 9 × 0	24. 3 × 7
25. 8 × 4	26. 3 × 3	27. 5 × 5	28. 2 × 6	29. 9 × 4	30. 5 × 3
31. 7 × 4	32. 4 × 4	33. 3 × 7	34. 9 × 2	35. 5 × 8	36. 10 × 5
37. 6 × 6	38. 7 × 5	39. 4 × 4	40. 9 × 3	41. 0 × 6	42. 10 × 4
43. 2 × 3	44. 3 × 3	45. 5 × 4	46. 6 × 7	47. 9 × 3	48. 2 × 7
49. 3 × 8	50. 5 × 5	51. 3 × 5	52. 6 × 2	53. 9 × 0	54. 5 × 10
55. 3 × 7	56. 9 × 3	57. 4 × 2	58. 5 × 0	59. 7 × 3	60. 5 × 4

Our New Fact Is  7

7 Facts
7 x 0 = 0
7 x 1 = 7
7 x 2 = 14
7 x 3 = 21
7 x 4 = 28
7 x 5 = 35
7 x 6 = 42
7 x 7 = 49
7 x 8 = 56
7 x 9 = 63
7 x 10 = 70

Draw straight lines from the facts to the correct answers!

7 x 2	70
7 x 7	21
7 x 6	28
7 x 0	63
7 x 5	14
7 x 1	35
7 x 10	56
7 x 4	0
7 x 8	7
7 x 9	42
7 x 3	49

Write Your 7 Facts

Trace it	Answer it	Fill in the blanks	Fill in the blanks	Write the fact
7 x 8 = 56	7 x 8 =	x 8 =	x =	
7 x 3 = 21	7 x 3 =	7 x =	x =	
7 x 10 = 70	7 x 10 =	x 10 =	x =	
7 x 2 = 14	7 x 2 =	7 x =	x =	
7 x 9 = 63	7 x 9 =	x 9 =	x =	
7 x 7 = 49	7 x 7 =	x 7 =	x =	
7 x 1 = 7	7 x 1 =	7 x =	x =	
7 x 6 = 42	7 x 6 =	7 x =	x =	
7 x 5 = 35	7 x 5 =	x 5 =	x =	
7 x 0 = 0	7 x 0 =	x 0 =	x =	
7 x 4 = 28	7 x 4 =	7 x =	x =	

Let's Practice!

Let's make sure you have all your facts down! Answer each multiplication problem.

$$7 \times 4$$

$$7 \times 9$$

$$7 \times 5$$

$$7 \times 6$$

$$7 \times 1$$

$$7 \times 7$$

$$7 \times 8$$

$$7 \times 0$$

$$7 \times 2$$

$$7 \times 10$$

$$7 \times 3$$

I can multiply by 7: TOP or BOTTOM

Name: _____ Date: _____

Goal: _____ problems in _____ seconds/minutes

1. 7 × 9	2. 7 × 6	3. 7 × 2	4. 7 × 7	5. 7 × 0	6. 7 × 10
7. 7 × 2	8. 7 × 8	9. 7 × 6	10. 7 × 4	11. 7 × 5	12. 7 × 8
13. 7 × 1	14. 7 × 8	15. 7 × 7	16. 7 × 7	17. 7 × 9	18. 7 × 1
19. 7 × 3	20. 7 × 0	21. 7 × 10	22. 7 × 2	23. 7 × 7	24. 7 × 9
25. 7 × 5	26. 7 × 4	27. 7 × 6	28. 7 × 3	29. 7 × 1	30. 7 × 0

31. 10 × 7	32. 2 × 7	33. 8 × 7	34. 4 × 7	35. 3 × 7	36. 8 × 7
37. 5 × 7	38. 8 × 7	39. 9 × 7	40. 4 × 7	41. 5 × 7	42. 1 × 7
43. 0 × 7	44. 8 × 7	45. 3 × 7	46. 10 × 7	47. 4 × 7	48. 8 × 7
49. 8 × 7	50. 3 × 7	51. 5 × 7	52. 6 × 7	53. 7 × 7	54. 2 × 7
55. 1 × 7	56. 9 × 7	57. 7 × 7	58. 6 × 7	59. 8 × 7	60. 0 × 7

I can multiply by 7s: Top AND Bottom

1. 7 x 8	2. 4 x 7	3. 2 x 7	4. 7 x 8	5. 0 x 7	6. 7 x 10
7. 2 x 7	8. 5 x 7	9. 7 x 8	10. 7 x 6	11. 2 x 7	12. 7 x 9
13. 4 x 7	14. 7 x 5	15. 7 x 0	16. 7 x 10	17. 9 x 7	18. 4 x 7
19. 7 x 2	20. 7 x 8	21. 9 x 7	22. 2 x 7	23. 7 x 7	24. 7 x 4
25. 7 x 3	26. 3 x 7	27. 7 x 6	28. 9 x 7	29. 7 x 1	30. 10 x 7
31. 2 x 7	32. 9 x 7	33. 7 x 7	34. 4 x 7	35. 3 x 7	36. 7 x 2
37. 7 x 7	38. 7 x 2	39. 7 x 4	40. 7 x 3	41. 10 x 7	42. 0 x 7
43. 7 x 1	44. 9 x 7	45. 8 x 7	46. 3 x 7	47. 7 x 2	48. 5 x 7
49. 4 x 7	50. 7 x 9	51. 7 x 7	52. 7 x 6	53. 5 x 7	54. 7 x 3
55. 7 x 2	56. 0 x 7	57. 2 x 7	58. 7 x 1	59. 8 x 7	60. 7 x 4

REVIEW: 0-7

1. 0×0	2. 8×6	3. 7×3	4. 9×7	5. 6×3	6. 2×0
7. 10×7	8. 3×6	9. 5×7	10. 9×2	11. 4×7	12. 3×8
13. 7×7	14. 9×7	15. 7×2	16. 4×7	17. 9×4	18. 3×10
19. 4×9	20. 0×4	21. 9×7	22. 10×7	23. 5×3	24. 2×7
25. 4×7	26. 2×8	27. 0×3	28. 4×10	29. 8×7	30. 4×5
31. 3×8	32. 9×7	33. 1×0	34. 9×6	35. 3×7	36. 3×3
37. 2×9	38. 0×7	39. 4×7	40. 6×2	41. 5×6	42. 0×5
43. 7×5	44. 9×7	45. 1×6	46. 3×9	47. 0×4	48. 7×5
49. 6×7	50. 8×7	51. 2×6	52. 0×5	53. 3×6	54. 8×4
55. 3×9	56. 10×7	57. 5×4	58. 8×3	59. 4×5	60. 2×9

REVIEW: 0-7

1. 6 x 6	2. 9 x 7	3. 3 x 5	4. 6 x 6	5. 0 x 8	6. 7 x 1
7. 2 x 5	8. 8 x 7	9. 5 x 5	10. 6 x 4	11. 7 x 9	12. 2 x 1
13. 4 x 5	14. 4 x 4	15. 7 x 2	16. 7 x 7	17. 0 x 1	18. 8 x 4
19. 2 x 0	20. 5 x 4	21. 3 x 4	22. 6 x 9	23. 0 x 3	24. 10 x 3
25. 4 x 4	26. 7 x 6	27. 0 x 2	28. 7 x 6	29. 4 x 8	30. 4 x 8
31. 2 x 6	32. 4 x 7	33. 8 x 7	34. 10 x 4	35. 3 x 5	36. 6 x 2
37. 3 x 6	38. 8 x 3	39. 7 x 5	40. 5 x 9	41. 2 x 10	42. 8 x 4
43. 3 x 6	44. 7 x 5	45. 6 x 3	46. 2 x 9	47. 6 x 6	48. 4 x 0
49. 3 x 4	50. 5 x 2	51. 8 x 1	52. 1 x 7	53. 6 x 3	54. 5 x 8
55. 3 x 3	56. 6 x 7	57. 4 x 9	58. 3 x 7	59. 6 x 5	60. 2 x 7

REVIEW: 0-7

1.	5 × 5	2.	0 × 3	3.	10 × 7	4.	4 × 6	5.	5 × 2	6.	9 × 5
7.	3 × 6	8.	9 × 4	9.	7 × 9	10.	2 × 7	11.	6 × 3	12.	1 × 1
13.	6 × 9	14.	3 × 7	15.	0 × 3	16.	10 × 5	17.	4 × 4	18.	3 × 7
19.	9 × 2	20.	5 × 7	21.	3 × 9	22.	10 × 4	23.	2 × 8	24.	6 × 1
25.	4 × 4	26.	8 × 3	27.	6 × 9	28.	0 × 1	29.	5 × 6	30.	3 × 9
31.	2 × 7	32.	6 × 6	33.	4 × 9	34.	3 × 10	35.	5 × 3	36.	8 × 7
37.	3 × 3	38.	7 × 5	39.	4 × 4	40.	9 × 2	41.	4 × 7	42.	7 × 7
43.	2 × 6	44.	5 × 3	45.	4 × 9	46.	3 × 9	47.	5 × 5	48.	0 × 3
49.	2 × 1	50.	7 × 5	51.	9 × 3	52.	5 × 7	53.	1 × 9	54.	0 × 9
55.	8 × 5	56.	2 × 7	57.	10 × 6	58.	3 × 8	59.	6 × 4	60.	2 × 8

Our New Fact Is 8

8 Facts
8 x 0 = 0
8 x 1 = 8
8 x 2 = 16
8 x 3 = 24
8 x 4 = 32
8 x 5 = 40
8 x 6 = 48
8 x 7 = 56
8 x 8 = 64
8 x 9 = 72
8 x 10 = 80

Draw dashed lines from the facts to the correct answers!

Fact	Answer
8 x 0	48
8 x 9	40
1 x 4	80
8 x 4	32
8 x 8	72
8 x 1	16
8 x 5	56
8 x 6	8
8 x 2	24
8 x 3	64
8 x 10	0

Write Your 8 Facts

Trace it	Answer it	Fill in the blanks	Fill in the blanks	Write the fact
8 x 4 = 32	8 x 4 =	x 4 =	x =	
8 x 0 = 0	8 x 0 =	8 x =	x =	
8 x 10 = 80	8 x 10 =	8 x =	x =	
8 x 5 = 40	8 x 5 =	x 5 =	x =	
8 x 9 = 72	8 x 9 =	x 9 =	x =	
8 x 1 = 8	8 x 1 =	8 x =	x =	
8 x 6 = 48	8 x 6 =	x 6 =	x =	
8 x 2 = 16	8 x 2 =	8 x =	x =	
8 x 8 = 64	8 x 8 =	x 8 =	x =	
8 x 3 = 24	8 x 3 =	8 x =	x =	
8 x 7 = 56	8 x 7 =	8 x =	x =	

Let's Practice!

Let's make sure you have all your facts down! Answer each multiplication problem.

$$\begin{array}{r} 8 \\ \times\ 8 \\ \hline \end{array}$$

$$\begin{array}{r} 8 \\ \times\ 7 \\ \hline \end{array}$$

$$\begin{array}{r} 8 \\ \times\ 9 \\ \hline \end{array}$$

$$\begin{array}{r} 8 \\ \times\ 3 \\ \hline \end{array}$$

$$\begin{array}{r} 8 \\ \times\ 4 \\ \hline \end{array}$$

$$\begin{array}{r} 8 \\ \times\ 1 \\ \hline \end{array}$$

$$\begin{array}{r} 8 \\ \times\ 10 \\ \hline \end{array}$$

$$\begin{array}{r} 8 \\ \times\ 2 \\ \hline \end{array}$$

$$\begin{array}{r} 8 \\ \times\ 6 \\ \hline \end{array}$$

$$\begin{array}{r} 8 \\ \times\ 0 \\ \hline \end{array}$$

$$\begin{array}{r} 8 \\ \times\ 5 \\ \hline \end{array}$$

I can multiply by 8: TOP or BOTTOM

Name: _____ Date: _____

Goal: _____ problems in _____ seconds/minutes

1. $\begin{array}{r} 8 \\ \times\ 0 \\ \hline \end{array}$	2. $\begin{array}{r} 8 \\ \times\ 2 \\ \hline \end{array}$	3. $\begin{array}{r} 8 \\ \times\ 1 \\ \hline \end{array}$	4. $\begin{array}{r} 8 \\ \times\ 10 \\ \hline \end{array}$	5. $\begin{array}{r} 8 \\ \times\ 8 \\ \hline \end{array}$	6. $\begin{array}{r} 8 \\ \times\ 6 \\ \hline \end{array}$
7. $\begin{array}{r} 8 \\ \times\ 6 \\ \hline \end{array}$	8. $\begin{array}{r} 8 \\ \times\ 3 \\ \hline \end{array}$	9. $\begin{array}{r} 8 \\ \times\ 2 \\ \hline \end{array}$	10. $\begin{array}{r} 8 \\ \times\ 8 \\ \hline \end{array}$	11. $\begin{array}{r} 8 \\ \times\ 9 \\ \hline \end{array}$	12. $\begin{array}{r} 8 \\ \times\ 4 \\ \hline \end{array}$
13. $\begin{array}{r} 8 \\ \times\ 5 \\ \hline \end{array}$	14. $\begin{array}{r} 8 \\ \times\ 7 \\ \hline \end{array}$	15. $\begin{array}{r} 8 \\ \times\ 6 \\ \hline \end{array}$	16. $\begin{array}{r} 8 \\ \times\ 1 \\ \hline \end{array}$	17. $\begin{array}{r} 8 \\ \times\ 9 \\ \hline \end{array}$	18. $\begin{array}{r} 8 \\ \times\ 1 \\ \hline \end{array}$
19. $\begin{array}{r} 8 \\ \times\ 3 \\ \hline \end{array}$	20. $\begin{array}{r} 8 \\ \times\ 2 \\ \hline \end{array}$	21. $\begin{array}{r} 8 \\ \times\ 3 \\ \hline \end{array}$	22. $\begin{array}{r} 8 \\ \times\ 5 \\ \hline \end{array}$	23. $\begin{array}{r} 8 \\ \times\ 8 \\ \hline \end{array}$	24. $\begin{array}{r} 8 \\ \times\ 9 \\ \hline \end{array}$
25. $\begin{array}{r} 8 \\ \times\ 7 \\ \hline \end{array}$	26. $\begin{array}{r} 8 \\ \times\ 3 \\ \hline \end{array}$	27. $\begin{array}{r} 8 \\ \times\ 8 \\ \hline \end{array}$	28. $\begin{array}{r} 8 \\ \times\ 0 \\ \hline \end{array}$	29. $\begin{array}{r} 8 \\ \times\ 1 \\ \hline \end{array}$	30. $\begin{array}{r} 8 \\ \times\ 10 \\ \hline \end{array}$

31. $\begin{array}{r} 7 \\ \times\ 8 \\ \hline \end{array}$	32. $\begin{array}{r} 3 \\ \times\ 8 \\ \hline \end{array}$	33. $\begin{array}{r} 4 \\ \times\ 8 \\ \hline \end{array}$	34. $\begin{array}{r} 5 \\ \times\ 8 \\ \hline \end{array}$	35. $\begin{array}{r} 2 \\ \times\ 8 \\ \hline \end{array}$	36. $\begin{array}{r} 10 \\ \times\ 8 \\ \hline \end{array}$
37. $\begin{array}{r} 0 \\ \times\ 8 \\ \hline \end{array}$	38. $\begin{array}{r} 3 \\ \times\ 8 \\ \hline \end{array}$	39. $\begin{array}{r} 5 \\ \times\ 8 \\ \hline \end{array}$	40. $\begin{array}{r} 7 \\ \times\ 8 \\ \hline \end{array}$	41. $\begin{array}{r} 9 \\ \times\ 8 \\ \hline \end{array}$	42. $\begin{array}{r} 4 \\ \times\ 8 \\ \hline \end{array}$
43. $\begin{array}{r} 2 \\ \times\ 8 \\ \hline \end{array}$	44. $\begin{array}{r} 1 \\ \times\ 8 \\ \hline \end{array}$	45. $\begin{array}{r} 10 \\ \times\ 8 \\ \hline \end{array}$	46. $\begin{array}{r} 4 \\ \times\ 8 \\ \hline \end{array}$	47. $\begin{array}{r} 3 \\ \times\ 8 \\ \hline \end{array}$	48. $\begin{array}{r} 2 \\ \times\ 8 \\ \hline \end{array}$
49. $\begin{array}{r} 8 \\ \times\ 8 \\ \hline \end{array}$	50. $\begin{array}{r} 7 \\ \times\ 8 \\ \hline \end{array}$	51. $\begin{array}{r} 6 \\ \times\ 8 \\ \hline \end{array}$	52. $\begin{array}{r} 5 \\ \times\ 8 \\ \hline \end{array}$	53. $\begin{array}{r} 9 \\ \times\ 8 \\ \hline \end{array}$	54. $\begin{array}{r} 0 \\ \times\ 8 \\ \hline \end{array}$
55. $\begin{array}{r} 10 \\ \times\ 8 \\ \hline \end{array}$	56. $\begin{array}{r} 2 \\ \times\ 8 \\ \hline \end{array}$	57. $\begin{array}{r} 8 \\ \times\ 8 \\ \hline \end{array}$	58. $\begin{array}{r} 3 \\ \times\ 8 \\ \hline \end{array}$	59. $\begin{array}{r} 5 \\ \times\ 8 \\ \hline \end{array}$	60. $\begin{array}{r} 2 \\ \times\ 8 \\ \hline \end{array}$

I can multiply by 8s: Top AND Bottom

Name: _____ Date: _____

Goal: _____ problems in _____ seconds/minutes

1.	7 x 8	2.	8 x 7	3.	8 x 2	4.	8 x 8	5.	9 x 8	6.	8 x 0
7.	8 x 1	8.	8 x 2	9.	6 x 8	10.	5 x 8	11.	8 x 9	12.	8 x 5
13.	9 x 8	14.	3 x 8	15.	8 x 7	16.	9 x 8	17.	8 x 1	18.	10 x 8
19.	0 x 8	20.	8 x 8	21.	7 x 8	22.	8 x 3	23.	8 x 2	24.	8 x 1
25.	8 x 8	26.	8 x 7	27.	8 x 3	28.	5 x 8	29.	8 x 9	30.	0 x 8
31.	10 x 8	32.	2 x 8	33.	8 x 4	34.	5 x 8	35.	8 x 3	36.	6 x 8
37.	8 x 2	38.	8 x 8	39.	9 x 8	40.	8 x 6	41.	5 x 8	42.	8 x 1
43.	8 x 0	44.	8 x 8	45.	8 x 8	46.	8 x 0	47.	8 x 10	48.	2 x 8
49.	3 x 8	50.	8 x 7	51.	8 x 6	52.	3 x 8	53.	4 x 8	54.	8 x 3
55.	7 x 8	56.	5 x 8	57.	8 x 9	58.	3 x 8	59.	8 x 4	60.	8 x 5

Name: _____ Date: _____

Goal: _____ problems in _____ seconds/minutes

1. 4 x 8	2. 9 x 8	3. 3 x 6	4. 2 x 8	5. 0 x 1	6. 8 x 8
7. 3 x 7	8. 6 x 4	9. 5 x 8	10. 2 x 0	11. 8 x 10	12. 6 x 2
13. 5 x 5	14. 6 x 8	15. 0 x 2	16. 1 x 7	17. 8 x 9	18. 3 x 2
19. 10 x 7	20. 3 x 4	21. 2 x 5	22. 3 x 3	23. 7 x 5	24. 1 x 9
25. 8 x 2	26. 3 x 6	27. 1 x 0	28. 4 x 8	29. 6 x 7	30. 9 x 3
31. 3 x 7	32. 8 x 4	33. 2 x 9	34. 0 x 7	35. 10 x 5	36. 3 x 2
37. 8 x 9	38. 6 x 4	39. 5 x 7	40. 2 x 9	41. 7 x 4	42. 5 x 1
43. 4 x 8	44. 7 x 5	45. 4 x 7	46. 2 x 9	47. 3 x 10	48. 4 x 3
49. 3 x 5	50. 6 x 2	51. 7 x 9	52. 10 x 7	53. 6 x 4	54. 2 x 6
55. 3 x 3	56. 6 x 5	57. 4 x 5	58. 8 x 3	59. 1 x 5	60. 9 x 8

Name: _____ Date: _____

Goal: _____ problems in _____ seconds/minutes

1. $\begin{array}{r} 4 \\ \times\ 7 \\ \hline \end{array}$	2. $\begin{array}{r} 8 \\ \times\ 6 \\ \hline \end{array}$	3. $\begin{array}{r} 4 \\ \times\ 5 \\ \hline \end{array}$	4. $\begin{array}{r} 1 \\ \times\ 0 \\ \hline \end{array}$	5. $\begin{array}{r} 5 \\ \times\ 10 \\ \hline \end{array}$	6. $\begin{array}{r} 5 \\ \times\ 4 \\ \hline \end{array}$
7. $\begin{array}{r} 6 \\ \times\ 9 \\ \hline \end{array}$	8. $\begin{array}{r} 0 \\ \times\ 3 \\ \hline \end{array}$	9. $\begin{array}{r} 1 \\ \times\ 4 \\ \hline \end{array}$	10. $\begin{array}{r} 6 \\ \times\ 2 \\ \hline \end{array}$	11. $\begin{array}{r} 8 \\ \times\ 7 \\ \hline \end{array}$	12. $\begin{array}{r} 5 \\ \times\ 6 \\ \hline \end{array}$
13. $\begin{array}{r} 3 \\ \times\ 5 \\ \hline \end{array}$	14. $\begin{array}{r} 7 \\ \times\ 2 \\ \hline \end{array}$	15. $\begin{array}{r} 8 \\ \times\ 7 \\ \hline \end{array}$	16. $\begin{array}{r} 5 \\ \times\ 8 \\ \hline \end{array}$	17. $\begin{array}{r} 2 \\ \times\ 1 \\ \hline \end{array}$	18. $\begin{array}{r} 9 \\ \times\ 7 \\ \hline \end{array}$
19. $\begin{array}{r} 8 \\ \times\ 10 \\ \hline \end{array}$	20. $\begin{array}{r} 7 \\ \times\ 5 \\ \hline \end{array}$	21. $\begin{array}{r} 3 \\ \times\ 7 \\ \hline \end{array}$	22. $\begin{array}{r} 1 \\ \times\ 9 \\ \hline \end{array}$	23. $\begin{array}{r} 0 \\ \times\ 5 \\ \hline \end{array}$	24. $\begin{array}{r} 2 \\ \times\ 7 \\ \hline \end{array}$
25. $\begin{array}{r} 8 \\ \times\ 8 \\ \hline \end{array}$	26. $\begin{array}{r} 4 \\ \times\ 5 \\ \hline \end{array}$	27. $\begin{array}{r} 2 \\ \times\ 9 \\ \hline \end{array}$	28. $\begin{array}{r} 7 \\ \times\ 5 \\ \hline \end{array}$	29. $\begin{array}{r} 3 \\ \times\ 8 \\ \hline \end{array}$	30. $\begin{array}{r} 0 \\ \times\ 5 \\ \hline \end{array}$
31. $\begin{array}{r} 2 \\ \times\ 5 \\ \hline \end{array}$	32. $\begin{array}{r} 4 \\ \times\ 6 \\ \hline \end{array}$	33. $\begin{array}{r} 9 \\ \times\ 3 \\ \hline \end{array}$	34. $\begin{array}{r} 10 \\ \times\ 8 \\ \hline \end{array}$	35. $\begin{array}{r} 7 \\ \times\ 5 \\ \hline \end{array}$	36. $\begin{array}{r} 2 \\ \times\ 4 \\ \hline \end{array}$
37. $\begin{array}{r} 3 \\ \times\ 5 \\ \hline \end{array}$	38. $\begin{array}{r} 2 \\ \times\ 8 \\ \hline \end{array}$	39. $\begin{array}{r} 9 \\ \times\ 7 \\ \hline \end{array}$	40. $\begin{array}{r} 6 \\ \times\ 4 \\ \hline \end{array}$	41. $\begin{array}{r} 10 \\ \times\ 4 \\ \hline \end{array}$	42. $\begin{array}{r} 2 \\ \times\ 4 \\ \hline \end{array}$
43. $\begin{array}{r} 7 \\ \times\ 5 \\ \hline \end{array}$	44. $\begin{array}{r} 4 \\ \times\ 6 \\ \hline \end{array}$	45. $\begin{array}{r} 9 \\ \times\ 4 \\ \hline \end{array}$	46. $\begin{array}{r} 2 \\ \times\ 5 \\ \hline \end{array}$	47. $\begin{array}{r} 4 \\ \times\ 4 \\ \hline \end{array}$	48. $\begin{array}{r} 2 \\ \times\ 7 \\ \hline \end{array}$
49. $\begin{array}{r} 3 \\ \times\ 9 \\ \hline \end{array}$	50. $\begin{array}{r} 8 \\ \times\ 7 \\ \hline \end{array}$	51. $\begin{array}{r} 5 \\ \times\ 7 \\ \hline \end{array}$	52. $\begin{array}{r} 3 \\ \times\ 0 \\ \hline \end{array}$	53. $\begin{array}{r} 2 \\ \times\ 7 \\ \hline \end{array}$	54. $\begin{array}{r} 8 \\ \times\ 4 \\ \hline \end{array}$
55. $\begin{array}{r} 1 \\ \times\ 2 \\ \hline \end{array}$	56. $\begin{array}{r} 5 \\ \times\ 3 \\ \hline \end{array}$	57. $\begin{array}{r} 2 \\ \times\ 6 \\ \hline \end{array}$	58. $\begin{array}{r} 8 \\ \times\ 4 \\ \hline \end{array}$	59. $\begin{array}{r} 2 \\ \times\ 4 \\ \hline \end{array}$	60. $\begin{array}{r} 5 \\ \times\ 5 \\ \hline \end{array}$

Name: _____ Date: _____

Goal: _____ problems in _____ seconds/minutes

1. $\begin{array}{r} 3 \\ \times\ 6 \\ \hline \end{array}$	2. $\begin{array}{r} 7 \\ \times\ 5 \\ \hline \end{array}$	3. $\begin{array}{r} 8 \\ \times\ 3 \\ \hline \end{array}$	4. $\begin{array}{r} 2 \\ \times\ 6 \\ \hline \end{array}$	5. $\begin{array}{r} 10 \\ \times\ 7 \\ \hline \end{array}$	6. $\begin{array}{r} 6 \\ \times\ 4 \\ \hline \end{array}$
7. $\begin{array}{r} 3 \\ \times\ 4 \\ \hline \end{array}$	8. $\begin{array}{r} 5 \\ \times\ 7 \\ \hline \end{array}$	9. $\begin{array}{r} 3 \\ \times\ 5 \\ \hline \end{array}$	10. $\begin{array}{r} 2 \\ \times\ 9 \\ \hline \end{array}$	11. $\begin{array}{r} 8 \\ \times\ 6 \\ \hline \end{array}$	12. $\begin{array}{r} 0 \\ \times\ 5 \\ \hline \end{array}$
13. $\begin{array}{r} 2 \\ \times\ 5 \\ \hline \end{array}$	14. $\begin{array}{r} 6 \\ \times\ 5 \\ \hline \end{array}$	15. $\begin{array}{r} 9 \\ \times\ 3 \\ \hline \end{array}$	16. $\begin{array}{r} 7 \\ \times\ 8 \\ \hline \end{array}$	17. $\begin{array}{r} 9 \\ \times\ 8 \\ \hline \end{array}$	18. $\begin{array}{r} 2 \\ \times\ 2 \\ \hline \end{array}$
19. $\begin{array}{r} 1 \\ \times\ 1 \\ \hline \end{array}$	20. $\begin{array}{r} 5 \\ \times\ 3 \\ \hline \end{array}$	21. $\begin{array}{r} 6 \\ \times\ 7 \\ \hline \end{array}$	22. $\begin{array}{r} 5 \\ \times\ 3 \\ \hline \end{array}$	23. $\begin{array}{r} 2 \\ \times\ 0 \\ \hline \end{array}$	24. $\begin{array}{r} 4 \\ \times\ 10 \\ \hline \end{array}$
25. $\begin{array}{r} 5 \\ \times\ 6 \\ \hline \end{array}$	26. $\begin{array}{r} 8 \\ \times\ 4 \\ \hline \end{array}$	27. $\begin{array}{r} 2 \\ \times\ 6 \\ \hline \end{array}$	28. $\begin{array}{r} 5 \\ \times\ 9 \\ \hline \end{array}$	29. $\begin{array}{r} 0 \\ \times\ 6 \\ \hline \end{array}$	30. $\begin{array}{r} 3 \\ \times\ 10 \\ \hline \end{array}$
31. $\begin{array}{r} 8 \\ \times\ 8 \\ \hline \end{array}$	32. $\begin{array}{r} 5 \\ \times\ 7 \\ \hline \end{array}$	33. $\begin{array}{r} 3 \\ \times\ 4 \\ \hline \end{array}$	34. $\begin{array}{r} 2 \\ \times\ 6 \\ \hline \end{array}$	35. $\begin{array}{r} 9 \\ \times\ 5 \\ \hline \end{array}$	36. $\begin{array}{r} 3 \\ \times\ 5 \\ \hline \end{array}$
37. $\begin{array}{r} 4 \\ \times\ 6 \\ \hline \end{array}$	38. $\begin{array}{r} 7 \\ \times\ 7 \\ \hline \end{array}$	39. $\begin{array}{r} 5 \\ \times\ 3 \\ \hline \end{array}$	40. $\begin{array}{r} 9 \\ \times\ 7 \\ \hline \end{array}$	41. $\begin{array}{r} 6 \\ \times\ 2 \\ \hline \end{array}$	42. $\begin{array}{r} 1 \\ \times\ 4 \\ \hline \end{array}$
43. $\begin{array}{r} 4 \\ \times\ 3 \\ \hline \end{array}$	44. $\begin{array}{r} 5 \\ \times\ 2 \\ \hline \end{array}$	45. $\begin{array}{r} 7 \\ \times\ 5 \\ \hline \end{array}$	46. $\begin{array}{r} 3 \\ \times\ 8 \\ \hline \end{array}$	47. $\begin{array}{r} 1 \\ \times\ 5 \\ \hline \end{array}$	48. $\begin{array}{r} 6 \\ \times\ 2 \\ \hline \end{array}$
49. $\begin{array}{r} 8 \\ \times\ 6 \\ \hline \end{array}$	50. $\begin{array}{r} 4 \\ \times\ 6 \\ \hline \end{array}$	51. $\begin{array}{r} 2 \\ \times\ 9 \\ \hline \end{array}$	52. $\begin{array}{r} 0 \\ \times\ 5 \\ \hline \end{array}$	53. $\begin{array}{r} 10 \\ \times\ 6 \\ \hline \end{array}$	54. $\begin{array}{r} 3 \\ \times\ 3 \\ \hline \end{array}$
55. $\begin{array}{r} 5 \\ \times\ 7 \\ \hline \end{array}$	56. $\begin{array}{r} 2 \\ \times\ 9 \\ \hline \end{array}$	57. $\begin{array}{r} 3 \\ \times\ 10 \\ \hline \end{array}$	58. $\begin{array}{r} 0 \\ \times\ 0 \\ \hline \end{array}$	59. $\begin{array}{r} 5 \\ \times\ 6 \\ \hline \end{array}$	60. $\begin{array}{r} 3 \\ \times\ 8 \\ \hline \end{array}$

Our New Fact Is 9

9 Facts
9 x 0 = 0
9 x 1 = 9
9 x 2 = 18
9 x 3 = 27
9 x 4 = 37
9 x 5 = 45
9 x 6 = 54
9 x 7 = 63
9 x 8 = 72
9 x 9 = 81
9 x 10 = 90

Draw squiggly lines from the facts to the correct answers!

9 x 5	0
9 x 9	9
9 x 6	81
9 x 0	18
9 x 1	72
9 x 7	27
9 x 2	36
9 x 10	90
9 x 3	63
9 x 4	45
9 x 8	54

Write Your 9 Facts

Trace it	Answer it	Fill in the blanks	Fill in the blanks	Write the fact
9 x 5 = 45	9 x 5 =	x 5 =	x =	
9 x 6 = 54	9 x 6 =	9 x =	x =	
9 x 1 = 9	9 x 1 =	x 1 =	x =	
9 x 4 = 36	9 x 4 =	9 x =	x =	
9 x 0 = 0	9 x 0 =	x 0 =	x =	
9 x 7 = 63	9 x 7 =	x 7 =	x =	
9 x 3 = 27	9 x 3 =	x 3 =	x =	
9 x 2 = 18	9 x 2 =	9 x =	x =	
9 x 8 = 72	9 x 8 =	9 x =	x =	
9 x 10 = 90	9 x 10 =	x 10 =	x =	
9 x 9 = 81	9 x 9 =	x 9 =	x =	

Let's Practice!

Let's make sure you have all your facts down! Answer each multiplication problem.

$$9 \times 10$$

$$9 \times 6$$

$$9 \times 5$$

$$9 \times 3$$

$$9 \times 7$$

$$9 \times 9$$

$$9 \times 1$$

$$9 \times 4$$

$$9 \times 8$$

$$9 \times 2$$

$$9 \times 0$$

I can multiply by 9: TOP or BOTTOM

1. 9 × 0	2. 9 × 8	3. 9 × 2	4. 9 × 4	5. 9 × 6	6. 9 × 3
7. 9 × 7	8. 9 × 4	9. 9 × 3	10. 9 × 7	11. 9 × 0	12. 9 × 10
13. 9 × 3	14. 9 × 8	15. 9 × 5	16. 9 × 2	17. 9 × 9	18. 9 × 10
19. 9 × 8	20. 9 × 5	21. 9 × 0	22. 9 × 2	23. 9 × 5	24. 9 × 7
25. 9 × 7	26. 9 × 3	27. 9 × 5	28. 9 × 8	29. 9 × 7	30. 9 × 10

―――――――――――――――――――――――――――――――――――――――

31. 10 × 9	32. 2 × 9	33. 8 × 9	34. 9 × 9	35. 5 × 9	36. 6 × 9
37. 3 × 9	38. 4 × 9	39. 9 × 9	40. 7 × 9	41. 8 × 9	42. 0 × 9
43. 2 × 9	44. 10 × 9	45. 3 × 9	46. 5 × 9	47. 8 × 9	48. 6 × 9
49. 4 × 9	50. 2 × 9	51. 4 × 9	52. 1 × 9	53. 0 × 9	54. 6 × 9
55. 2 × 9	56. 7 × 9	57. 10 × 9	58. 0 × 9	59. 4 × 9	60. 3 × 9

I can multiply by 9s: Top AND Bottom

Name: _____ Date: _____

Goal: _____ problems in _____ seconds/minutes

1. $\begin{array}{r} 3 \\ \times\ 9 \\ \hline \end{array}$
2. $\begin{array}{r} 9 \\ \times\ 2 \\ \hline \end{array}$
3. $\begin{array}{r} 9 \\ \times\ 9 \\ \hline \end{array}$
4. $\begin{array}{r} 9 \\ \times\ 0 \\ \hline \end{array}$
5. $\begin{array}{r} 10 \\ \times\ 9 \\ \hline \end{array}$
6. $\begin{array}{r} 9 \\ \times\ 3 \\ \hline \end{array}$

7. $\begin{array}{r} 9 \\ \times\ 5 \\ \hline \end{array}$
8. $\begin{array}{r} 9 \\ \times\ 3 \\ \hline \end{array}$
9. $\begin{array}{r} 8 \\ \times\ 9 \\ \hline \end{array}$
10. $\begin{array}{r} 9 \\ \times\ 5 \\ \hline \end{array}$
11. $\begin{array}{r} 6 \\ \times\ 9 \\ \hline \end{array}$
12. $\begin{array}{r} 8 \\ \times\ 9 \\ \hline \end{array}$

13. $\begin{array}{r} 9 \\ \times\ 10 \\ \hline \end{array}$
14. $\begin{array}{r} 8 \\ \times\ 9 \\ \hline \end{array}$
15. $\begin{array}{r} 0 \\ \times\ 9 \\ \hline \end{array}$
16. $\begin{array}{r} 1 \\ \times\ 9 \\ \hline \end{array}$
17. $\begin{array}{r} 9 \\ \times\ 2 \\ \hline \end{array}$
18. $\begin{array}{r} 4 \\ \times\ 9 \\ \hline \end{array}$

19. $\begin{array}{r} 3 \\ \times\ 9 \\ \hline \end{array}$
20. $\begin{array}{r} 7 \\ \times\ 9 \\ \hline \end{array}$
21. $\begin{array}{r} 9 \\ \times\ 8 \\ \hline \end{array}$
22. $\begin{array}{r} 4 \\ \times\ 9 \\ \hline \end{array}$
23. $\begin{array}{r} 9 \\ \times\ 8 \\ \hline \end{array}$
24. $\begin{array}{r} 9 \\ \times\ 0 \\ \hline \end{array}$

25. $\begin{array}{r} 9 \\ \times\ 1 \\ \hline \end{array}$
26. $\begin{array}{r} 9 \\ \times\ 2 \\ \hline \end{array}$
27. $\begin{array}{r} 9 \\ \times\ 9 \\ \hline \end{array}$
28. $\begin{array}{r} 9 \\ \times\ 6 \\ \hline \end{array}$
29. $\begin{array}{r} 5 \\ \times\ 9 \\ \hline \end{array}$
30. $\begin{array}{r} 4 \\ \times\ 9 \\ \hline \end{array}$

31. $\begin{array}{r} 3 \\ \times\ 9 \\ \hline \end{array}$
32. $\begin{array}{r} 9 \\ \times\ 6 \\ \hline \end{array}$
33. $\begin{array}{r} 9 \\ \times\ 9 \\ \hline \end{array}$
34. $\begin{array}{r} 7 \\ \times\ 9 \\ \hline \end{array}$
35. $\begin{array}{r} 9 \\ \times\ 4 \\ \hline \end{array}$
36. $\begin{array}{r} 3 \\ \times\ 9 \\ \hline \end{array}$

37. $\begin{array}{r} 9 \\ \times\ 10 \\ \hline \end{array}$
38. $\begin{array}{r} 9 \\ \times\ 9 \\ \hline \end{array}$
39. $\begin{array}{r} 9 \\ \times\ 3 \\ \hline \end{array}$
40. $\begin{array}{r} 9 \\ \times\ 2 \\ \hline \end{array}$
41. $\begin{array}{r} 9 \\ \times\ 5 \\ \hline \end{array}$
42. $\begin{array}{r} 10 \\ \times\ 9 \\ \hline \end{array}$

43. $\begin{array}{r} 4 \\ \times\ 9 \\ \hline \end{array}$
44. $\begin{array}{r} 7 \\ \times\ 9 \\ \hline \end{array}$
45. $\begin{array}{r} 9 \\ \times\ 8 \\ \hline \end{array}$
46. $\begin{array}{r} 8 \\ \times\ 9 \\ \hline \end{array}$
47. $\begin{array}{r} 9 \\ \times\ 9 \\ \hline \end{array}$
48. $\begin{array}{r} 0 \\ \times\ 9 \\ \hline \end{array}$

49. $\begin{array}{r} 9 \\ \times\ 2 \\ \hline \end{array}$
50. $\begin{array}{r} 9 \\ \times\ 6 \\ \hline \end{array}$
51. $\begin{array}{r} 1 \\ \times\ 9 \\ \hline \end{array}$
52. $\begin{array}{r} 9 \\ \times\ 10 \\ \hline \end{array}$
53. $\begin{array}{r} 7 \\ \times\ 9 \\ \hline \end{array}$
54. $\begin{array}{r} 9 \\ \times\ 4 \\ \hline \end{array}$

55. $\begin{array}{r} 3 \\ \times\ 9 \\ \hline \end{array}$
56. $\begin{array}{r} 9 \\ \times\ 2 \\ \hline \end{array}$
57. $\begin{array}{r} 9 \\ \times\ 8 \\ \hline \end{array}$
58. $\begin{array}{r} 6 \\ \times\ 9 \\ \hline \end{array}$
59. $\begin{array}{r} 4 \\ \times\ 9 \\ \hline \end{array}$
60. $\begin{array}{r} 9 \\ \times\ 7 \\ \hline \end{array}$

Name: _____ Date: _____

Goal: _____ problems in _____ seconds/minutes

1. 7 x 9	2. 4 x 6	3. 9 x 9	4. 1 x 5	5. 4 x 7	6. 7 x 0
7. 4 x 7	8. 9 x 1	9. 3 x 5	10. 6 x 2	11. 9 x 8	12. 5 x 5
13. 2 x 5	14. 7 x 7	15. 0 x 4	16. 9 x 6	17. 3 x 7	18. 1 x 6
19. 9 x 9	20. 5 x 3	21. 7 x 5	22. 6 x 9	23. 2 x 0	24. 3 x 6
25. 4 x 9	26. 5 x 9	27. 0 x 2	28. 9 x 10	29. 5 x 3	30. 2 x 6
31. 9 x 4	32. 7 x 2	33. 0 x 7	34. 5 x 3	35. 4 x 8	36. 10 x 5
37. 3 x 6	38. 7 x 2	39. 9 x 5	40. 7 x 8	41. 3 x 9	42. 2 x 0
43. 0 x 5	44. 4 x 6	45. 8 x 9	46. 9 x 10	47. 4 x 2	48. 5 x 4
49. 8 x 5	50. 9 x 3	51. 1 x 8	52. 6 x 4	53. 6 x 6	54. 6 x 2
55. 0 x 9	56. 4 x 2	57. 1 x 6	58. 8 x 7	59. 4 x 9	60. 5 x 6

REVIEW: 0-9

Name: _____ Date: _____

Goal: _____ problems in _____ seconds/minutes

1. 3 × 5	2. 6 × 7	3. 2 × 8	4. 0 × 2	5. 7 × 10	6. 6 × 8
7. 9 × 2	8. 6 × 5	9. 3 × 7	10. 0 × 5	11. 6 × 2	12. 1 × 8
13. 6 × 6	14. 8 × 3	15. 7 × 7	16. 5 × 2	17. 3 × 9	18. 0 × 0
19. 2 × 6	20. 5 × 4	21. 7 × 8	22. 2 × 6	23. 9 × 5	24. 2 × 4
25. 8 × 4	26. 5 × 6	27. 3 × 8	28. 9 × 9	29. 4 × 7	30. 6 × 2
31. 8 × 7	32. 4 × 6	33. 5 × 7	34. 9 × 4	35. 2 × 6	36. 4 × 4
37. 3 × 7	38. 8 × 4	39. 3 × 6	40. 5 × 5	41. 9 × 8	42. 2 × 5
43. 5 × 6	44. 4 × 4	45. 8 × 2	46. 0 × 8	47. 5 × 0	48. 4 × 7
49. 9 × 2	50. 6 × 4	51. 7 × 3	52. 2 × 5	53. 9 × 7	54. 6 × 7
55. 2 × 4	56. 5 × 2	57. 8 × 4	58. 6 × 5	59. 3 × 9	60. 3 × 10

REVIEW: 0-9

Name: _____ Date: _____

Goal: _____ problems in _____ seconds/minutes

1. 8 x 5	2. 3 x 2	3. 6 x 9	4. 4 x 0	5. 7 x 3	6. 4 x 4
7. 3 x 5	8. 2 x 2	9. 6 x 7	10. 8 x 3	11. 9 x 5	12. 9 x 9
13. 10 x 8	14. 5 x 6	15. 4 x 7	16. 8 x 8	17. 3 x 2	18. 6 x 5
19. 5 x 6	20. 3 x 3	21. 7 x 8	22. 4 x 9	23. 8 x 2	24. 1 x 3
25. 7 x 7	26. 4 x 3	27. 5 x 6	28. 9 x 4	29. 10 x 2	30. 5 x 4
31. 4 x 6	32. 5 x 5	33. 7 x 2	34. 1 x 9	35. 5 x 10	36. 6 x 4
37. 2 x 3	38. 5 x 3	39. 6 x 2	40. 5 x 4	41. 8 x 8	42. 7 x 6
43. 5 x 6	44. 2 x 7	45. 8 x 7	46. 6 x 5	47. 4 x 7	48. 9 x 9
49. 3 x 5	50. 4 x 2	51. 1 x 2	52. 3 x 8	53. 7 x 6	54. 4 x 9
55. 0 x 5	56. 3 x 4	57. 2 x 7	58. 8 x 6	59. 5 x 3	60. 6 x 5

Our New Fact Is

10 Facts
10 x 0 = 0
10 x 1 = 10
10 x 2 = 20
10 x 3 = 30
10 x 4 = 40
10 x 5 = 50
10 x 6 = 60
10 x 7 = 70
10 x 8 = 80
10 x 9 = 90
10 x 10 = 100

Draw straight lines from the facts to the correct answers!

10 x 4	100
10 x 0	10
10 x 5	90
10 x 2	0
10 x 7	50
10 x 10	60
10 x 6	20
10 x 1	80
10 x 9	40
10 x 3	30
10 x 8	70

Write Your 10 Facts

Trace it	Answer it	Fill in the blanks	Fill in the blanks	Write the fact
10 x 2 = 20	10 x 2 =	10 x ___ =	___ x ___ =	
10 x 7 = 70	10 x 7 =	___ x 7 =	___ x ___ =	
10 x 3 = 30	10 x 3 =	10 x ___ =	___ x ___ =	
10 x 6 = 60	10 x 6 =	10 x ___ =	___ x ___ =	
10 x 8 = 80	10 x 8 =	10 x ___ =	___ x ___ =	
10 x 10 = 100	10 x 10 =	___ x 10 =	___ x ___ =	
10 x 0 = 0	10 x 0 =	___ x 0 =	___ x ___ =	
10 x 5 = 50	10 x 5 =	10 x ___ =	___ x ___ =	
10 x 9 = 90	10 x 9 =	10 x ___ =	___ x ___ =	
10 x 1 = 10	10 x 1 =	___ x 1 =	___ x ___ =	
10 x 4 = 40	10 x 4 =	10 x ___ =	___ x ___ =	

Let's Practice!

Let's make sure you have all your facts down! Answer each multiplication problem.

$$10 \times 3$$

$$10 \times 8$$

$$10 \times 6$$

$$10 \times 10$$

$$10 \times 2$$

$$10 \times 7$$

$$10 \times 5$$

$$10 \times 0$$

$$10 \times 1$$

$$10 \times 4$$

$$10 \times 9$$

I can multiply by 10: TOP or BOTTOM

1. 10 × 4	2. 10 × 3	3. 10 × 8	4. 10 × 4	5. 10 × 10	6. 10 × 2
7. 10 × 5	8. 10 × 8	9. 10 × 5	10. 10 × 2	11. 10 × 3	12. 10 × 9
13. 10 × 7	14. 10 × 8	15. 10 × 4	16. 10 × 9	17. 10 × 2	18. 10 × 0
19. 10 × 1	20. 10 × 0	21. 10 × 5	22. 10 × 10	23. 10 × 3	24. 10 × 2
25. 10 × 9	26. 10 × 4	27. 10 × 7	28. 10 × 5	29. 10 × 2	30. 10 × 8

---

31. 9 × 10	32. 4 × 10	33. 7 × 10	34. 9 × 10	35. 2 × 10	36. 0 × 10
37. 10 × 10	38. 4 × 10	39. 7 × 10	40. 4 × 10	41. 3 × 10	42. 2 × 10
43. 8 × 10	44. 4 × 10	45. 7 × 10	46. 6 × 10	47. 2 × 10	48. 0 × 10
49. 2 × 10	50. 6 × 10	51. 4 × 10	52. 8 × 10	53. 5 × 10	54. 9 × 10
55. 10 × 10	56. 9 × 10	57. 2 × 10	58. 3 × 10	59. 8 × 10	60. 5 × 10

I can multiply by 10s: Top AND Bottom

1. 10
    x 8

2. 10
    x 10

3.   3
    x 10

4. 10
    x 8

5.   4
    x 10

6. 10
    x 2

7.   9
    x 10

8.   4
    x 10

9. 10
    x 3

10. 10
    x 9

11. 0
    x 10

12. 10
    x 1

13. 7
    x 10

14. 10
    x 7

15. 10
    x 3

16. 10
    x 9

17. 5
    x 10

18. 2
    x 10

19. 10
    x 9

20. 10
    x 3

21. 2
    x 10

22. 8
    x 10

23. 10
    x 9

24. 10
    x 4

25. 10
    x 2

26. 7
    x 10

27. 10
    x 9

28. 10
    x 10

29. 10
    x 0

30. 4
    x 10

31. 9
    x 10

32. 2
    x 10

33. 10
    x 8

34. 4
    x 10

35. 5
    x 10

36. 10
    x 8

37. 1
    x 10

38. 10
    x 8

39. 10
    x 3

40. 10
    x 9

41. 0
    x 10

42. 2
    x 10

43. 10
    x 7

44. 3
    x 10

45. 7
    x 10

46. 8
    x 10

47. 10
    x 9

48. 2
    x 10

49. 2
    x 10

50. 10
    x 10

51. 5
    x 10

52. 10
    x 0

53. 2
    x 10

54. 10
    x 4

55. 10
    x 9

56. 3
    x 10

57. 4
    x 10

58. 10
    x 7

59. 5
    x 10

60. 10
    x 10

REVIEW: 0-10

1. 3×10	2. 7×6	3. 2×1	4. 0×8	5. 7×10	6. 10×10
7. 6×7	8. 3×7	9. 8×8	10. 9×2	11. 10×8	12. 6×5
13. 2×4	14. 5×3	15. 6×6	16. 8×3	17. 10×7	18. 4×4
19. 0×9	20. 3×7	21. 6×6	22. 4×2	23. 9×10	24. 4×2
25. 4×6	26. 10×10	27. 8×6	28. 5×2	29. 4×3	30. 8×2
31. 2×2	32. 5×4	33. 6×3	34. 4×9	35. 10×3	36. 3×5
37. 4×6	38. 7×2	39. 9×4	40. 2×8	41. 9×9	42. 2×4
43. 2×3	44. 6×5	45. 8×3	46. 2×6	47. 7×5	48. 0×6
49. 3×10	50. 8×5	51. 2×6	52. 7×5	53. 7×7	54. 2×5
55. 1×3	56. 4×10	57. 9×7	58. 5×6	59. 2×4	60. 5×2

REVIEW: 0-10

Name: _____ Date: _____

Goal: _____ problems in _____ seconds/minutes

1. 3 x 4	2. 6 x 2	3. 1 x 8	4. 9 x 5	5. 3 x 6	6. 1 x 0
7. 10 x 7	8. 6 x 5	9. 4 x 7	10. 8 x 2	11. 3 x 3	12. 9 x 9
13. 7 x 5	14. 3 x 8	15. 0 x 5	16. 3 x 10	17. 9 x 8	18. 3 x 3
19. 4 x 3	20. 5 x 7	21. 8 x 3	22. 2 x 9	23. 0 x 4	24. 2 x 3
25. 3 x 5	26. 6 x 2	27. 7 x 7	28. 8 x 4	29. 3 x 7	30. 1 x 9
31. 0 x 10	32. 6 x 4	33. 3 x 6	34. 7 x 4	35. 5 x 5	36. 2 x 9
37. 4 x 7	38. 9 x 9	39. 7 x 4	40. 2 x 6	41. 5 x 7	42. 9 x 4
43. 4 x 5	44. 2 x 6	45. 8 x 8	46. 4 x 9	47. 3 x 1	48. 5 x 3
49. 7 x 7	50. 4 x 3	51. 2 x 8	52. 9 x 10	53. 5 x 6	54. 7 x 3
55. 8 x 4	56. 7 x 6	57. 3 x 6	58. 5 x 5	59. 0 x 5	60. 2 x 2

REVIEW: 0-10

Name: _____ Date: _____

Goal: _____ problems in _____ seconds/minutes

1. 4 x 6	2. 2 x 7	3. 10 x 9	4. 10 x 10	5. 7 x 4	6. 2 x 3
7. 5 x 7	8. 7 x 2	9. 8 x 6	10. 3 x 5	11. 2 x 9	12. 6 x 0
13. 5 x 6	14. 7 x 2	15. 8 x 8	16. 7 x 4	17. 2 x 9	18. 10 x 8
19. 4 x 3	20. 2 x 6	21. 8 x 4	22. 5 x 5	23. 2 x 7	24. 6 x 3
25. 5 x 7	26. 8 x 8	27. 6 x 4	28. 2 x 9	29. 3 x 7	30. 6 x 4
31. 3 x 6	32. 7 x 7	33. 8 x 3	34. 1 x 5	35. 4 x 7	36. 7 x 5
37. 5 x 2	38. 6 x 5	39. 3 x 7	40. 9 x 5	41. 2 x 7	42. 6 x 6
43. 8 x 5	44. 6 x 7	45. 8 x 2	46. 6 x 4	47. 5 x 3	48. 3 x 10
49. 2 x 4	50. 10 x 10	51. 8 x 6	52. 5 x 7	53. 2 x 8	54. 9 x 7
55. 9 x 8	56. 5 x 4	57. 6 x 6	58. 7 x 8	59. 2 x 0	60. 3 x 1

Page 6

Our New Fact Is 0

0 Facts
0 x 0 = 0
0 x 1 = 0
0 x 2 = 0
0 x 3 = 0
0 x 4 = 0
0 x 5 = 0
0 x 6 = 0
0 x 7 = 0
0 x 8 = 0
0 x 9 = 0
0 x 10 = 0

Draw squiggly lines from the facts to the correct answers!

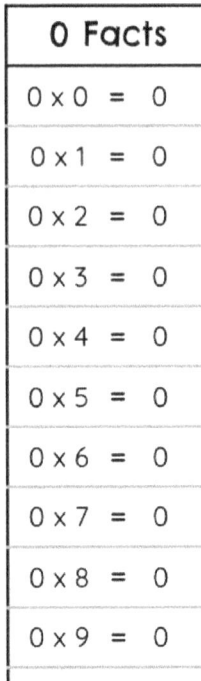

0 x 0 — 0
0 x 4 — 0
0 x 10 — 0
0 x 3 — 0
0 x 2 — 0
0 x 9 — 0
0 x 1 — 0
0 x 7 — 0
0 x 8 — 0
0 x 1 — 0
0 x 6 — 0

Page 7

Write Your 0 Facts

Trace it	Answer it	Fill in the blanks	Fill in the blanks	Write the fact
0 x 3 = 0	0 x 3 = 0	0 x 3 = 0	0 x 3 = 0	0 x 3 = 0
0 x 1 = 0	0 x 1 = 0	0 x 1 = 0	0 x 1 = 0	0 x 1 = 0
0 x 8 = 0	0 x 8 = 0	0 x 8 = 0	0 x 8 = 0	0 x 8 = 0
0 x 0 = 0	0 x 0 = 0	0 x 0 = 0	0 x 0 = 0	0 x 0 = 0
0 x 4 = 0	0 x 4 = 0	0 x 4 = 0	0 x 4 = 0	0 x 4 = 0
0 x 2 = 0	0 x 2 = 0	0 x 2 = 0	0 x 2 = 0	0 x 2 = 0
0 x 10 = 0	0 x 10 = 0	0 x 10 = 0	0 x 10 = 0	0 x 10 = 0
0 x 7 = 0	0 x 7 = 0	0 x 7 = 0	0 x 7 = 0	0 x 7 = 0
0 x 5 = 0	0 x 5 = 0	0 x 5 = 0	0 x 5 = 0	0 x 5 = 0
0 x 9 = 0	0 x 9 = 0	0 x 9 = 0	0 x 9 = 0	0 x 9 = 0
0 x 6 = 0	0 x 6 = 0	0 x 6 = 0	0 x 6 = 0	0 x 6 = 0

Page 8

Let's Practice!

Let's make sure you have all your facts down! Answer each multiplication problem.

$$\begin{array}{r} 0 \\ \times\ 6 \\ \hline 0 \end{array}$$

$$\begin{array}{r} 0 \\ \times\ 10 \\ \hline 0 \end{array}$$

$$\begin{array}{r} 0 \\ \times\ 2 \\ \hline 0 \end{array}$$

$$\begin{array}{r} 0 \\ \times\ 4 \\ \hline 0 \end{array}$$

$$\begin{array}{r} 0 \\ \times\ 7 \\ \hline 0 \end{array}$$

$$\begin{array}{r} 0 \\ \times\ 3 \\ \hline 0 \end{array}$$

$$\begin{array}{r} 0 \\ \times\ 8 \\ \hline 0 \end{array}$$

$$\begin{array}{r} 0 \\ \times\ 1 \\ \hline 0 \end{array}$$

$$\begin{array}{r} 0 \\ \times\ 9 \\ \hline 0 \end{array}$$

$$\begin{array}{r} 0 \\ \times\ 5 \\ \hline 0 \end{array}$$

$$\begin{array}{r} 0 \\ \times\ 0 \\ \hline 0 \end{array}$$

Page 9

I can multiply by 0: TOP, BOTTOM AND MIXED

Name: _____ Date: _____
Goal: _____ problems in _____ seconds/minutes

0 × 3 = 0	0 × 8 = 0	0 × 3 = 0	0 × 2 = 0	0 × 5 = 0	0 × 4 = 0
0 × 7 = 0	0 × 2 = 0	0 × 6 = 0	0 × 0 = 0	0 × 10 = 0	0 × 1 = 0
0 × 1 = 0	0 × 6 = 0	0 × 4 = 0	0 × 9 = 0	0 × 5 = 0	0 × 7 = 0
8 × 0 = 0	10 × 0 = 0	8 × 0 = 0	1 × 0 = 0	3 × 0 = 0	0 × 0 = 0
4 × 0 = 0	1 × 0 = 0	2 × 0 = 0	10 × 0 = 0	4 × 0 = 0	6 × 0 = 0
2 × 0 = 0	9 × 0 = 0	7 × 0 = 0	9 × 0 = 0	3 × 0 = 0	5 × 0 = 0
10 × 0 = 0	3 × 0 = 0	0 × 6 = 0	0 × 2 = 0	9 × 0 = 0	0 × 5 = 0
0 × 1 = 0	0 × 5 = 0	9 × 0 = 0	4 × 0 = 0	0 × 0 = 0	0 × 2 = 0
6 × 0 = 0	0 × 3 = 0	0 × 0 = 0	0 × 7 = 0	1 × 0 = 0	8 × 0 = 0
0 × 1 = 0	0 × 3 = 0	2 × 0 = 0	0 × 8 = 0	0 × 10 = 0	10 × 0 = 0

Page 10

Our New Fact Is ①

1 Facts
1 x 0 = 0
1 x 1 = 1
1 x 2 = 2
1 x 3 = 3
1 x 4 = 4
1 x 5 = 5
1 x 6 = 6
1 x 7 = 7
1 x 8 = 8
1 x 9 = 9
1 x 10 = 10

Draw straight lines from the facts to the correct answers!

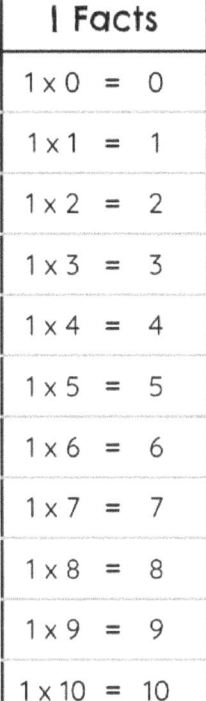

Page 11

Write Your 1 Facts

Trace it	Answer it	Fill in the blanks	Fill in the blanks	Write the fact
1 x 2 = 2	1 x 2 = 2	1 x 2 = 2	1 x 2 = 2	1 x 2 = 2
1 x 7 = 7	1 x 7 = 7	1 x 7 = 7	1 x 7 = 7	1 x 7 = 7
1 x 3 = 3	1 x 3 = 3	1 x 3 = 3	1 x 3 = 3	1 x 3 = 3
1 x 6 = 6	1 x 6 = 6	1 x 6 = 6	1 x 6 = 6	1 x 6 = 6
1 x 8 = 8	1 x 8 = 8	1 x 8 = 8	1 x 8 = 8	1 x 8 = 8
1 x 10 = 10	1 x 10 = 10	1 x 10 = 10	1 x 10 = 10	1 x 10 = 10
1 x 0 = 0	1 x 0 = 0	1 x 0 = 0	1 x 0 = 0	1 x 0 = 0
1 x 5 = 5	1 x 5 = 5	1 x 5 = 5	1 x 5 = 5	1 x 5 = 5
1 x 9 = 9	1 x 9 = 9	1 x 9 = 9	1 x 9 = 9	1 x 9 = 9
1 x 1 = 1	1 x 1 = 1	1 x 1 = 1	1 x 1 = 1	1 x 1 = 1
1 x 4 = 4	1 x 4 = 4	1 x 4 = 4	1 x 4 = 4	1 x 4 = 4

Page 12

Let's Practice!

Let's make sure you have all your facts down! Answer each multiplication problem

$$\begin{array}{r} 1 \\ \times\ 4 \\ \hline 4 \end{array}\qquad \begin{array}{r} 1 \\ \times\ 7 \\ \hline 7 \end{array}\qquad \begin{array}{r} 1 \\ \times\ 2 \\ \hline 2 \end{array}$$

$$\begin{array}{r} 1 \\ \times\ 9 \\ \hline 9 \end{array}\qquad \begin{array}{r} 1 \\ \times\ 5 \\ \hline 5 \end{array}\qquad \begin{array}{r} 1 \\ \times\ 8 \\ \hline 8 \end{array}$$

$$\begin{array}{r} 1 \\ \times\ 0 \\ \hline 0 \end{array}$$

$$\begin{array}{r} 1 \\ \times\ 3 \\ \hline 3 \end{array}\qquad \begin{array}{r} 1 \\ \times\ 6 \\ \hline 6 \end{array}\qquad \begin{array}{r} 1 \\ \times\ 1 \\ \hline 1 \end{array}$$

$$\begin{array}{r} 1 \\ \times\ 10 \\ \hline 10 \end{array}$$

Page 13

I can multiply by 1: TOP or BOTTOM	Name: _____ Date: _____
	Goal: ____ problems in ____ seconds/minutes

×2	×10	×7	×4	×3	×9
2	**10**	**7**	**4**	**3**	**9**
×1	×0	×6	×6	×5	×7
1	**0**	**6**	**6**	**5**	**7**
×10	×8	×9	×2	×1	×3
10	**8**	**9**	**2**	**1**	**3**
×4	×9	×2	×1	×0	×8
4	**9**	**2**	**1**	**0**	**8**
×3	×4	×7	×2	×1	×5
3	**4**	**7**	**2**	**1**	**5**
10 ×1	5 ×1	6 ×1	2 ×1	8 ×1	6 ×1
10	**5**	**6**	**2**	**8**	**6**
5 ×1	6 ×1	4 ×1	0 ×1	1 ×1	3 ×1
5	**6**	**4**	**0**	**1**	**3**
9 ×1	10 ×1	4 ×1	6 ×1	2 ×1	5 ×1
9	**10**	**4**	**6**	**2**	**5**
8 ×1	9 ×1	2 ×1	2 ×1	4 ×1	7 ×1
8	**9**	**2**	**2**	**4**	**7**
3 ×1	3 ×1	5 ×1	4 ×1	9 ×1	0 ×1
7	**3**	**5**	**4**	**9**	**0**

1 Facts Answer Keys

Page 14

I can multiply by 1s: Top AND Bottom

Name: _____ Date: _____
Goal: _____ problems in _____ seconds/minutes

1×5 = **5**	6×1 = **6**	9×1 = **9**	1×1 = **1**	0×1 = **0**	1×3 = **3**
6	**4**	**8**	**0**	**2**	**4**
3	**6**	**9**	**7**	**8**	**10**
8	**6**	**3**	**2**	**1**	**8**
0	**9**	**3**	**5**	**6**	**2**
0	**5**	**3**	**9**	**7**	**5**
4	**3**	**8**	**9**	**0**	**10**
2	**3**	**6**	**7**	**8**	**5**
4	**7**	**8**	**9**	**1**	**2**
3	**4**	**8**	**7**	**5**	**6**

Page 15

REVIEW: 0-1

Name: _____ Date: _____
Goal: _____ problems in _____ seconds/minutes

0	**2**	**7**	**4**	**0**	**0**
1	**7**	**0**	**0**	**2**	**8**
0	**3**	**0**	**0**	**2**	**8**
9	**0**	**3**	**9**	**0**	**0**
9	**1**	**0**	**0**	**0**	**0**
0	**0**	**10**	**2**	**8**	**0**
8	**2**	**0**	**0**	**0**	**6**
7	**3**	**0**	**0**	**0**	**0**
0	**7**	**0**	**8**	**5**	**0**
2	**0**	**0**	**3**	**0**	**6**

Page 16

REVIEW: 0-1

Name: _____ Date: _____
Goal: _____ problems in _____ seconds/minutes

4	**0**	**0**	**8**	**10**	**2**
0	**0**	**0**	**6**	**3**	**0**
10	**0**	**3**	**0**	**1**	**0**
0	**0**	**8**	**0**	**0**	**5**
0	**0**	**8**	**8**	**0**	**10**
8	**0**	**7**	**0**	**0**	**0**
5	**0**	**8**	**0**	**0**	**10**
0	**0**	**3**	**6**	**3**	**5**
0	**0**	**0**	**0**	**3**	**5**
7	**0**	**7**	**0**	**0**	**0**

Page 17

REVIEW: 0-1

Name: _____ Date: _____
Goal: _____ problems in _____ seconds/minutes

5	**0**	**7**	**0**	**0**	**4**
10	**0**	**9**	**4**	**0**	**0**
0	**7**	**0**	**6**	**0**	**3**
0	**0**	**2**	**0**	**0**	**0**
0	**0**	**6**	**7**	**0**	**0**
0	**0**	**5**	**1**	**7**	**0**
7	**0**	**0**	**3**	**0**	**2**
0	**3**	**0**	**8**	**6**	**8**
0	**0**	**0**	**8**	**0**	**0**
10	**0**	**7**	**0**	**2**	**8**

Page 18

Our New Fact Is ②

2 Facts
2 × 0 = 0
2 × 1 = 2
2 × 2 = 4
2 × 3 = 6
2 × 4 = 8
2 × 5 = 10
2 × 6 = 12
2 × 7 = 14
2 × 8 = 16
2 × 9 = 18
2 × 10 = 20

Draw dashed lines from the facts to the correct answers!

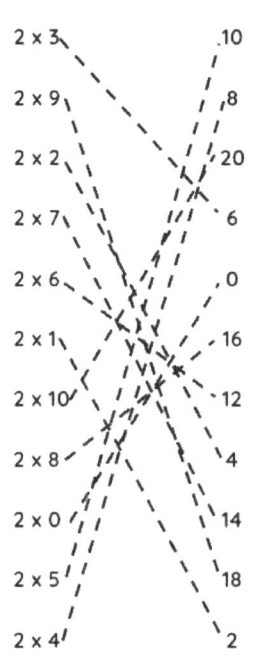

Page 19

Write Your 2 Facts

Trace it	Answer it	Fill in the blanks	Fill in the blanks	Write the fact
2 x 10 = 20	2 x 10 = 20	2 x 10 = 20	2 x 10 = 20	2 x 10 = 20
2 x 3 = 6	2 x 3 = 6	2 x 3 = 6	2 x 3 = 6	2 x 3 = 6
2 x 6 = 12	2 x 6 = 12	2 x 6 = 12	2 x 6 = 12	2 x 6 = 12
2 x 8 = 16	2 x 8 = 16	2 x 8 = 16	2 x 8 = 16	2 x 8 = 16
2 x 4 = 8	2 x 4 = 8	2 x 4 = 8	2 x 4 = 8	2 x 4 = 8
2 x 1 = 2	2 x 1 = 2	2 x 1 = 2	2 x 1 = 2	2 x 1 = 2
2 x 9 = 18	2 x 9 = 18	2 x 9 = 18	2 x 9 = 18	2 x 9 = 18
2 x 2 = 4	2 x 2 = 4	2 x 2 = 4	2 x 2 = 4	2 x 2 = 4
2 x 0 = 0	2 x 0 = 0	2 x 0 = 0	2 x 0 = 0	2 x 0 = 0
2 x 5 = 10	2 x 5 = 10	2 x 5 = 10	2 x 5 = 10	2 x 5 = 10
2 x 7 = 14	2 x 7 = 14	2 x 7 = 14	2 x 7 = 14	2 x 7 = 14

Page 20

Let's Practice!

Let's make sure you have all your facts down! Answer each multiplication problem.

$$\begin{array}{r} 2 \\ \times\ 8 \\ \hline 16 \end{array} \qquad \begin{array}{r} 2 \\ \times\ 3 \\ \hline 6 \end{array} \qquad \begin{array}{r} 2 \\ \times\ 2 \\ \hline 4 \end{array}$$

$$\begin{array}{r} 2 \\ \times\ 1 \\ \hline 2 \end{array} \qquad \begin{array}{r} 2 \\ \times\ 10 \\ \hline 20 \end{array} \qquad \begin{array}{r} 2 \\ \times\ 9 \\ \hline 18 \end{array}$$

$$\begin{array}{r} 2 \\ \times\ 7 \\ \hline 14 \end{array}$$

$$\begin{array}{r} 2 \\ \times\ 5 \\ \hline 10 \end{array} \qquad \begin{array}{r} 2 \\ \times\ 0 \\ \hline 0 \end{array}$$

$$\begin{array}{r} 2 \\ \times\ 4 \\ \hline 8 \end{array} \qquad \begin{array}{r} 2 \\ \times\ 6 \\ \hline 12 \end{array}$$

Page 21

I can multiply by 2:
TOP or BOTTOM

Name: _____ Date: _____
Goal: _____ problems in _____ seconds/minutes

2 × 2 = 4	2 × 1 = 2	2 × 8 = 16	2 × 10 = 20	2 × 9 = 18	2 × 5 = 10
2 × 3 = 6	2 × 7 = 14	2 × 6 = 12	2 × 8 = 16	2 × 1 = 2	2 × 0 = 0
2 × 9 = 18	2 × 8 = 16	2 × 6 = 12	2 × 5 = 10	2 × 4 = 8	2 × 8 = 16
2 × 1 = 2	2 × 2 = 4	2 × 0 = 0	2 × 10 = 20	2 × 9 = 18	2 × 3 = 6
2 × 1 = 2	2 × 2 = 4	2 × 7 = 14	2 × 7 = 14	2 × 5 = 10	2 × 3 = 6
5 × 2 = 10	4 × 2 = 8	0 × 2 = 0	10 × 2 = 20	9 × 2 = 18	6 × 2 = 12
7 × 2 = 14	4 × 2 = 8	3 × 2 = 6	2 × 2 = 4	1 × 2 = 2	0 × 2 = 0
7 × 2 = 14	9 × 2 = 18	3 × 2 = 6	4 × 2 = 8	5 × 2 = 10	5 × 2 = 10
7 × 2 = 14	2 × 2 = 4	1 × 2 = 2	8 × 2 = 16	6 × 2 = 12	9 × 2 = 18
10 × 2 = 20	2 × 2 = 4	4 × 2 = 8	7 × 2 = 14	9 × 2 = 18	3 × 2 = 6

2 Facts Answer Keys

Page 22

I can multiply by 2s: Top AND Bottom

Name: _____ Date: _____
Goal: ____ problems in ____ seconds/minutes

8	8	6	18	20	2
0	18	16	10	8	2
6	16	20	0	2	4
16	10	6	4	14	12
12	8	18	6	4	2
0	8	10	10	14	18
10	12	16	14	4	6
20	0	4	18	8	6
12	14	16	8	18	20
10	4	2	12	8	14

Page 23

I can multiply by 2s: Top AND Bottom

Name: _____ Date: _____
Goal: ____ problems in ____ seconds/minutes

8	8	6	18	20	2
0	18	16	10	8	2
6	16	20	0	2	4
16	10	6	4	14	12
12	8	18	6	4	2
0	8	10	10	14	18
10	12	16	14	4	6
20	0	4	18	8	6
12	14	16	8	18	20
10	4	2	12	8	14

Page 24

REVIEW: 0-2

Name: _____ Date: _____
Goal: ____ problems in ____ seconds/minutes

4	12	0	14	9	8
0	7	16	0	9	18
4	0	0	10	14	8
3	6	0	10	16	7
3	10	0	0	4	14
16	8	0	3	0	14
12	4	0	16	10	8
8	18	0	0	8	16
4	7	0	14	10	9
20	7	18	6	0	2

Page 25

REVIEW: 0-2

Name: _____ Date: _____
Goal: ____ problems in ____ seconds/minutes

7	0	4	14	0	18
0	6	0	20	1	18
8	6	0	16	0	3
0	0	4	8	6	12
8	0	0	3	12	0
12	1	14	0	8	4
2	0	3	10	0	6
8	6	10	18	9	0
12	7	0	6	20	9
14	5	8	16	3	18

Page 26

Our New Fact Is

3 Facts
3 x 0 = 0
3 x 1 = 3
3 x 2 = 6
3 x 3 = 9
3 x 4 = 12
3 x 5 = 15
3 x 6 = 18
3 x 7 = 21
3 x 8 = 24
3 x 9 = 27
3 x 10 = 30

Draw squiggly lines from the facts to the correct answers!

3 x 5	15
3 x 1	21
3 x 2	24
3 x 7	12
3 x 10	18
3 x 9	27
3 x 3	3
3 x 8	9
3 x 6	30
3 x 4	6
3 x 0	0

Page 27

Write Your 3 Facts

Trace it	Answer it	Fill in the blanks	Fill in the blanks	Write the fact
3 x 6 = 18	3 x 6 = 18	3 x 6 = 18	3 x 6 = 18	3 x 6 = 18
3 x 2 = 6	3 x 2 = 6	3 x 2 = 6	3 x 2 = 6	3 x 2 = 6
3 x 7 = 14	3 x 7 = 14	3 x 7 = 14	3 x 7 = 14	3 x 7 = 14
3 x 1 = 3	3 x 1 = 3	3 x 1 = 3	3 x 1 = 3	3 x 1 = 3
3 x 5 = 15	3 x 5 = 15	3 x 5 = 15	3 x 5 = 15	3 x 5 = 15
3 x 8 = 24	3 x 8 = 24	3 x 8 = 24	3 x 8 = 24	3 x 8 = 24
3 x 4 = 12	3 x 4 = 12	3 x 4 = 12	3 x 4 = 12	3 x 4 = 12
3 x 0 = 0	3 x 0 = 0	3 x 0 = 0	3 x 0 = 0	3 x 0 = 0
3 x 9 = 27	3 x 9 = 27	3 x 9 = 27	3 x 9 = 27	3 x 9 = 27
3 x 3 = 9	3 x 3 = 9	3 x 3 = 9	3 x 3 = 9	3 x 3 = 9
3 x 10 = 30	3 x 10 = 30	3 x 10 = 30	3 x 10 = 30	3 x 10 = 30

Page 28

Let's Practice!

Let's make sure you have all your facts down! Answer each multiplication problem.

$$\begin{array}{r} 3 \\ \times\ 3 \\ \hline 9 \end{array} \qquad \begin{array}{r} 3 \\ \times\ 7 \\ \hline 21 \end{array} \qquad \begin{array}{r} 3 \\ \times\ 10 \\ \hline 20 \end{array}$$

$$\begin{array}{r} 3 \\ \times\ 2 \\ \hline 6 \end{array} \qquad \begin{array}{r} 3 \\ \times\ 4 \\ \hline 12 \end{array} \qquad \begin{array}{r} 3 \\ \times\ 6 \\ \hline 18 \end{array}$$

$$\begin{array}{r} 3 \\ \times\ 5 \\ \hline 15 \end{array}$$

$$\begin{array}{r} 3 \\ \times\ 9 \\ \hline 27 \end{array} \qquad \begin{array}{r} 3 \\ \times\ 2 \\ \hline 6 \end{array} \qquad \begin{array}{r} 3 \\ \times\ 0 \\ \hline 0 \end{array}$$

$$\begin{array}{r} 3 \\ \times\ 8 \\ \hline 24 \end{array}$$

Page 29

I can multiply by 3: TOP or BOTTOM

Name: _____ Date: _____
Goal: ____ problems in ____ seconds/minutes

3 ×5	3 ×2	3 ×3	3 ×1	3 ×0	3 ×10
15	6	9	3	0	30
3 ×6	3 ×9	3 ×7	3 ×3	3 ×2	3 ×7
18	27	21	9	6	21
3 ×9	3 ×5	3 ×4	3 ×1	3 ×0	3 ×10
27	15	12	3	0	30
3 ×9	3 ×8	3 ×3	3 ×4	3 ×2	3 ×8
27	24	9	12	6	24
3 ×9	3 ×7	3 ×4	3 ×8	3 ×8	3 ×2
27	21	12	18	24	6
1 ×3	0 ×3	10 ×3	4 ×3	8 ×3	9 ×3
3	0	30	12	24	27
3 ×3	4 ×3	0 ×3	10 ×3	8 ×3	3 ×3
9	12	0	30	24	9
9 ×3	8 ×3	3 ×3	2 ×3	6 ×3	7 ×3
27	24	9	6	18	21
2 ×3	2 ×3	6 ×3	9 ×3	5 ×3	1 ×3
6	6	18	27	15	3
10 ×3	9 ×3	2 ×3	4 ×3	3 ×3	7 ×3
30	27	6	12	9	21

3 Facts Answer Keys

Page 30

I can multiply by 3s: Top AND Bottom

Name: _____ Date: _____
Goal: ____ problems in ____ seconds/minutes

9×3=**27**	3×1=**3**	1×3=**3**	3×4=**12**	6×3=**18**	3×9=**27**
3×4=**12**	3×3=**9**	8×3=**24**	3×9=**27**	7×3=**21**	4×3=**12**
3×1=**3**	0×3=**0**	10×3=**30**	9×3=**27**	3×2=**6**	4×3=**12**
5×3=**15**	7×3=**21**	3×1=**3**	8×3=**24**	3×5=**15**	3×1=**3**
3×0=**0**	3×10=**30**	8×3=**24**	3×6=**18**	9×3=**27**	3×3=**9**
1×3=**3**	3×0=**0**	4×3=**12**	3×3=**9**	3×8=**24**	7×3=**21**
3×4=**12**	5×3=**15**	3×2=**6**	3×8=**24**	3×7=**21**	4×3=**12**
1×3=**3**	0×3=**0**	3×10=**30**	5×3=**15**	3×3=**9**	7×3=**21**
3×9=**27**	3×6=**18**	2×3=**6**	3×1=**3**	0×3=**0**	3×10=**30**
7×3=**21**	3×2=**6**	3×1=**3**	0×3=**0**	3×3=**9**	3×5=**15**

Page 31

REVIEW: 0-3

Name: _____ Date: _____
Goal: ____ problems in ____ seconds/minutes

3×2=**6**	7×3=**21**	1×9=**9**	2×6=**12**	3×4=**12**	0×3=**0**
0×6=**0**	3×3=**9**	2×8=**15**	6×1=**6**	5×3=**15**	0×0=**0**
4×2=**8**	2×8=**16**	3×2=**6**	0×4=**0**	7×1=**7**	1×3=**3**
3×8=**24**	0×2=**0**	1×8=**8**	7×3=**21**	3×5=**15**	1×1=**1**
10×3=**30**	3×7=**21**	1×6=**6**	0×3=**6**	2×4=**8**	6×1=**6**
7×2=**14**	2×5=**10**	0×3=**0**	10×2=**20**	3×7=**21**	2×2=**4**
3×9=**27**	4×1=**4**	0×4=**0**	3×6=**18**	2×1=**2**	3×8=**24**
2×4=**8**	7×3=**21**	9×1=**9**	2×5=**10**	0×10=**0**	2×10=**20**
2×5=**10**	6×1=**6**	0×5=**0**	9×2=**18**	2×4=**8**	0×2=**0**
10×3=**30**	4×2=**8**	2×6=**12**	8×2=**16**	1×7=**7**	3×3=**9**

Page 32

REVIEW: 0-3

Name: _____ Date: _____
Goal: ____ problems in ____ seconds/minutes

6×2=**12**	7×1=**7**	0×3=**0**	2×2=**4**	1×0=**0**	10×1=**10**
3×7=**21**	1×5=**5**	8×1=**9**	9×0=**0**	2×4=**8**	0×3=**0**
1×4=**4**	3×6=**18**	8×2=**16**	9×3=**27**	1×10=**10**	0×4=**0**
5×0=**0**	10×3=**30**	6×2=**12**	3×4=**12**	3×8=**24**	2×0=**0**
7×1=**7**	5×0=**0**	2×7=**14**	8×2=**16**	1×0=**0**	2×2=**4**
4×3=**12**	3×3=**9**	2×8=**16**	4×0=**0**	0×6=**0**	10×2=**20**
2×9=**18**	4×1=**4**	0×7=**0**	10×3=**30**	9×3=**27**	3×9=**27**
2×5=**10**	5×3=**15**	0×9=**0**	6×0=**0**	1×3=**3**	3×8=**24**
0×4=**0**	2×8=**16**	1×8=**8**	9×3=**27**	2×3=**6**	5×1=**5**
10×0=**10**	0×6=**0**	3×5=**15**	2×8=**16**	1×4=**4**	3×2=**6**

Page 33

REVIEW: 0-3

Name: _____ Date: _____
Goal: ____ problems in ____ seconds/minutes

0×0=**0**	8×3=**24**	2×5=**10**	0×2=**0**	10×2=**20**	2×7=**14**
2×8=**16**	8×2=**16**	1×2=**2**	4×0=**0**	0×5=**0**	3×1=**3**
4×3=**12**	6×2=**12**	2×7=**14**	3×4=**12**	8×3=**24**	2×9=**18**
1×6=**6**	3×7=**21**	8×3=**24**	10×0=**0**	3×0=**0**	3×2=**6**
2×4=**8**	8×2=**16**	9×3=**27**	1×5=**5**	4×1=**4**	3×7=**21**
10×1=**10**	1×1=**1**	5×2=**10**	7×3=**21**	3×9=**27**	2×5=**10**
2×2=**4**	4×1=**4**	1×7=**7**	9×2=**18**	0×7=**0**	3×0=**0**
2×4=**8**	6×2=**12**	7×3=**21**	0×8=**0**	2×0=**0**	1×6=**6**
9×2=**18**	10×3=**30**	6×3=**18**	2×9=**18**	3×5=**15**	2×3=**6**
1×6=**6**	6×3=**18**	2×8=**16**	9×0=**0**	10×2=**20**	6×2=**12**

Page 34

Our New Fact Is

4 Facts	
4 x 0 =	0
4 x 1 =	4
4 x 2 =	8
4 x 3 =	12
4 x 4 =	16
4 x 5 =	20
4 x 6 =	24
4 x 7 =	28
4 x 8 =	32
4 x 9 =	36
4 x 10 =	40

Draw straight lines from the facts to the correct answers!

4 x 1 — 32
4 x 8 — 4
4 x 0 — 16
4 x 7 — 0
4 x 3 — 8
4 x 10 — 28
4 x 6 — 12
4 x 4 — 40
4 x 2 — 20
4 x 9 — 36
4 x 5 — 24

Page 35

Write Your 4 Facts

Trace it	Answer it	Fill in the blanks	Fill in the blanks	Write the fact
4 x 2 = 8	4 x 2 = 8	4 x 2 = 8	4 x 2 = 8	4 x 2 = 8
4 x 7 = 28	4 x 7 = 28	4 x 7 = 28	4 x 7 = 28	4 x 7 = 28
4 x 3 = 12	4 x 3 = 12	4 x 3 = 12	4 x 3 = 12	4 x 3 = 12
4 x 6 = 24	4 x 6 = 24	4 x 6 = 24	4 x 6 = 24	4 x 6 = 24
4 x 8 = 32	4 x 8 = 32	4 x 8 = 32	4 x 8 = 32	4 x 8 = 32
4 x 10 = 40	4 x 10 = 40	4 x 10 = 40	4 x 10 = 40	4 x 10 = 40
4 x 0 = 0	4 x 0 = 0	4 x 0 = 0	4 x 0 = 0	4 x 0 = 0
4 x 5 = 20	4 x 5 = 20	4 x 5 = 20	4 x 5 = 20	4 x 5 = 20
4 x 9 = 36	4 x 9 = 36	4 x 9 = 36	4 x 9 = 36	4 x 9 = 36
4 x 1 = 4	4 x 1 = 4	4 x 1 = 4	4 x 1 = 4	4 x 1 = 4
4 x 4 = 16	4 x 4 = 16	4 x 4 = 16	4 x 4 = 16	4 x 4 = 16

Page 36

Let's Practice!

Let's make sure you have all your facts down! Answer each multiplication problem.

$$\begin{array}{r} 4 \\ \times\ 5 \\ \hline 20 \end{array}$$

$$\begin{array}{r} 4 \\ \times\ 1 \\ \hline 4 \end{array}$$

$$\begin{array}{r} 4 \\ \times\ 8 \\ \hline 32 \end{array}$$

$$\begin{array}{r} 4 \\ \times\ 6 \\ \hline 24 \end{array}$$

$$\begin{array}{r} 4 \\ \times\ 2 \\ \hline 8 \end{array}$$

$$\begin{array}{r} 4 \\ \times\ 4 \\ \hline 16 \end{array}$$

$$\begin{array}{r} 4 \\ \times\ 9 \\ \hline 36 \end{array}$$

$$\begin{array}{r} 4 \\ \times\ 0 \\ \hline 0 \end{array}$$

$$\begin{array}{r} 4 \\ \times\ 10 \\ \hline 40 \end{array}$$

$$\begin{array}{r} 4 \\ \times\ 3 \\ \hline 12 \end{array}$$

$$\begin{array}{r} 4 \\ \times\ 7 \\ \hline 28 \end{array}$$

Page 37

I can multiply by 4: TOP or BOTTOM

Name: _____ Date: _____
Goal: _____ problems in _____ seconds/minutes

4 ×7	4 ×2	4 ×8	4 ×0	4 ×10	4 ×2
28	8	32	0	40	8

4 ×4	4 ×5	4 ×9	4 ×7	4 ×2	4 ×4
16	20	32	28	8	16

4 ×5	4 ×0	4 ×8	4 ×1	4 ×10	4 ×3
20	0	32	4	40	12

4 ×8	4 ×9	4 ×3	4 ×5	4 ×6	4 ×8
32	36	12	20	24	32

4 ×9	4 ×3	4 ×7	4 ×5	4 ×4	4 ×1
36	12	28	20	16	4

0 ×4	8 ×4	5 ×4	3 ×4	8 ×4	9 ×4
0	32	20	12	32	36

10 ×4	2 ×4	3 ×4	9 ×4	6 ×4	4 ×4
40	8	12	36	24	16

9 ×4	8 ×4	3 ×4	5 ×4	2 ×4	4 ×4
36	32	12	20	8	16

0 ×4	10 ×4	9 ×4	3 ×4	5 ×4	8 ×4
0	40	36	12	20	32

4 ×4	6 ×4	5 ×4	9 ×4	7 ×4	3 ×4
16	24	20	36	28	12

Page 38

I can multiply by 4s: Top AND Bottom

Name: _____ Date: _____
Goal: ____ problems in ____ seconds/minutes

4 × 8 = **32**	4 × 4 = **16**	3 × 4 = **12**	4 × 9 = **36**	0 × 4 = **0**	4 × 10 = **40**
9 × 4 = **36**	6 × 4 = **24**	4 × 8 = **32**	4 × 2 = **8**	1 × 4 = **4**	4 × 0 = **0**
2 × 4 = **8**	4 × 3 = **12**	4 × 7 = **28**	4 × 9 = **36**	8 × 4 = **32**	4 × 4 = **16**
4 × 3 = **12**	4 × 10 = **40**	5 × 4 = **20**	8 × 4 = **32**	4 × 8 = **32**	4 × 3 = **12**
4 × 9 = **36**	7 × 4 = **28**	4 × 3 = **12**	6 × 4 = **24**	4 × 7 = **28**	4 × 4 = **16**
8 × 4 = **32**	1 × 4 = **4**	4 × 0 = **0**	10 × 4 = **40**	8 × 4 = **32**	4 × 3 = **12**
4 × 4 = **16**	4 × 6 = **24**	4 × 5 = **20**	4 × 8 = **32**	4 × 4 = **28**	4 × 6 = **24**
4 × 6 = **24**	1 × 4 = **4**	10 × 4 = **40**	0 × 4 = **0**	4 × 2 = **8**	4 × 4 = **16**
3 × 4 = **12**	4 × 8 = **32**	6 × 4 = **24**	4 × 5 = **20**	7 × 4 = **28**	4 × 3 = **12**
4 × 9 = **36**	8 × 4 = **32**	5 × 4 = **20**	4 × 3 = **12**	1 × 4 = **4**	4 × 2 = **8**

Page 39

REVIEW: 0-4

Name: _____ Date: _____
Goal: ____ problems in ____ seconds/minutes

4 × 2 = **8**	6 × 3 = **18**	2 × 8 = **16**	4 × 9 = **36**	10 × 3 = **30**	0 × 4 = **0**
3 × 4 = **12**	7 × 4 = **28**	1 × 9 = **9**	0 × 3 = **0**	6 × 1 = **6**	1 × 5 = **5**
0 × 4 = **0**	3 × 7 = **21**	8 × 4 = **32**	4 × 10 = **40**	2 × 7 = **14**	9 × 3 = **27**
5 × 4 = **20**	4 × 9 = **36**	3 × 6 = **18**	1 × 9 = **9**	4 × 2 = **8**	0 × 3 = **0**
8 × 0 = **0**	6 × 2 = **12**	1 × 5 = **5**	0 × 5 = **0**	4 × 1 = **4**	9 × 2 = **18**
7 × 4 = **28**	3 × 4 = **12**	8 × 4 = **32**	2 × 9 = **18**	10 × 3 = **30**	7 × 2 = **14**
0 × 6 = **5**	4 × 0 = **0**	1 × 8 = **8**	9 × 2 = **0**	4 × 5 = **0**	7 × 1 = **10**
0 × 0 = **0**	4 × 4 = **16**	9 × 4 = **36**	2 × 7 = **14**	3 × 9 = **27**	1 × 0 = **0**
8 × 2 = **16**	4 × 6 = **24**	0 × 9 = **0**	4 × 10 = **40**	3 × 6 = **18**	2 × 4 = **8**
7 × 1 = **7**	9 × 3 = **27**	1 × 8 = **8**	0 × 3 = **0**	2 × 6 = **12**	4 × 8 = **32**

Page 40

REVIEW: 0-4

Name: _____ Date: _____
Goal: ____ problems in ____ seconds/minutes

7 × 2 = **14**	5 × 2 = **10**	2 × 9 = **18**	0 × 2 = **0**	1 × 7 = **7**	8 × 3 = **24**
4 × 4 = **16**	7 × 3 = **21**	4 × 8 = **32**	2 × 10 = **20**	3 × 7 = **21**	9 × 2 = **18**
6 × 4 = **24**	2 × 7 = **14**	3 × 9 = **27**	10 × 0 = **0**	3 × 3 = **9**	2 × 8 = **16**
5 × 3 = **15**	4 × 7 = **28**	2 × 8 = **16**	1 × 0 = **0**	2 × 6 = **12**	4 × 3 = **12**
9 × 3 = **27**	7 × 4 = **28**	2 × 2 = **4**	6 × 4 = **24**	0 × 9 = **0**	1 × 7 = **7**
1 × 1 = **1**	3 × 4 = **12**	9 × 4 = **36**	4 × 6 = **24**	2 × 0 = **0**	3 × 6 = **18**
10 × 3 = **30**	9 × 4 = **36**	2 × 6 = **12**	2 × 5 = **10**	0 × 5 = **0**	2 × 8 = **16**
7 × 2 = **14**	5 × 4 = **20**	1 × 8 = **8**	3 × 6 = **18**	10 × 4 = **40**	7 × 1 = **7**
0 × 3 = **0**	2 × 9 = **18**	4 × 0 = **0**	3 × 5 = **15**	8 × 4 = **32**	4 × 9 = **36**
9 × 4 = **36**	10 × 2 = **20**	8 × 1 = **8**	4 × 6 = **24**	7 × 2 = **14**	1 × 6 = **6**

Page 41

REVIEW: 0-4

Name: _____ Date: _____
Goal: ____ problems in ____ seconds/minutes

6 × 3 = **18**	7 × 2 = **14**	0 × 8 = **0**	10 × 4 = **40**	0 × 0 = **0**	3 × 3 = **9**
0 × 7 = **0**	5 × 3 = **15**	2 × 8 = **16**	4 × 4 = **16**	3 × 3 = **9**	9 × 0 = **0**
1 × 7 = **7**	3 × 9 = **27**	4 × 7 = **28**	9 × 4 = **36**	2 × 5 = **10**	1 × 8 = **8**
0 × 5 = **0**	3 × 0 = **0**	2 × 7 = **14**	4 × 9 = **36**	0 × 2 = **0**	4 × 4 = **16**
4 × 0 = **0**	5 × 3 = **15**	4 × 8 = **32**	0 × 10 = **0**	7 × 3 = **21**	2 × 6 = **12**
5 × 4 = **20**	4 × 5 = **20**	8 × 2 = **16**	6 × 1 = **6**	3 × 10 = **30**	7 × 0 = **0**
3 × 6 = **18**	9 × 3 = **27**	0 × 7 = **0**	1 × 5 = **5**	4 × 8 = **32**	9 × 2 = **18**
2 × 9 = **18**	5 × 2 = **9**	1 × 7 = **0**	0 × 5 = **40**	3 × 1 = **21**	9 × 3 = **4**
10 × 2 = **18**	6 × 0 = **10**	8 × 2 = **7**	0 × 7 = **0**	3 × 7 = **3**	8 × 4 = **27**
(10 × 2) = **20**	**0**	**16**	**0**	**21**	**32**

5 Facts Answer Keys

Page 42

Our New Fact Is

5 Facts
5 x 0 = 0
5 x 1 = 5
5 x 2 = 10
5 x 3 = 15
5 x 4 = 20
5 x 5 = 25
5 x 6 = 30
5 x 7 = 35
5 x 8 = 40
5 x 9 = 45
5 x 10 = 50

Draw dashed lines from the facts to the correct answers!

5 x 3	0
5 x 10	25
5 x 4	45
5 x 1	5
5 x 6	40
5 x 9	10
5 x 5	35
5 x 0	15
5 x 8	50
5 x 2	20
5 x 7	30

Page 43

Write Your 5 Facts

Trace it	Answer it	Fill in the blanks	Fill in the blanks	Write the fact
5 x 3 = 15	5 x 3 = 15	5 x 3 = 15	5 x 3 = 15	5 x 3 = 15
5 x 4 = 20	5 x 4 = 20	5 x 4 = 20	5 x 4 = 20	5 x 4 = 20
5 x 0 = 0	5 x 0 = 0	5 x 0 = 0	5 x 0 = 0	5 x 0 = 0
5 x 10 = 50	5 x 10 = 50	5 x 10 = 50	5 x 10 = 50	5 x 10 = 50
5 x 5 = 25	5 x 5 = 25	5 x 5 = 25	5 x 5 = 25	5 x 5 = 25
5 x 1 = 5	5 x 1 = 5	5 x 1 = 5	5 x 1 = 5	5 x 1 = 5
5 x 9 = 45	5 x 9 = 45	5 x 9 = 45	5 x 9 = 45	5 x 9 = 45
5 x 6 = 30	5 x 6 = 30	5 x 6 = 30	5 x 6 = 30	5 x 6 = 30
5 x 2 = 10	5 x 2 = 10	5 x 2 = 10	5 x 2 = 10	5 x 2 = 10
5 x 8 = 40	5 x 8 = 40	5 x 8 = 40	5 x 8 = 40	5 x 8 = 40
5 x 7 = 35	5 x 7 = 35	5 x 7 = 35	5 x 7 = 35	5 x 7 = 35

Page 44

Let's Practice!

Let's make sure you have all your facts down! Answer each multiplication problem.

$$5 \times 3 = 15$$

$$5 \times 1 = 5$$

$$5 \times 5 = 25$$

$$5 \times 0 = 0$$

$$5 \times 9 = 45$$

$$5 \times 8 = 40$$

$$5 \times 10 = 50$$

$$5 \times 6 = 30$$

$$5 \times 2 = 10$$

$$5 \times 4 = 20$$

$$5 \times 7 = 35$$

Page 45

I can multiply by 5: TOP or BOTTOM

Name: _____ Date: _____

Goal: ____ problems in ____ seconds/minutes

5 x 8 = 40	5 x 2 = 10	5 x 1 = 5	5 x 0 = 0	5 x 10 = 50	5 x 9 = 45
5 x 7 = 35	5 x 3 = 15	5 x 7 = 35	5 x 3 = 15	5 x 9 = 45	5 x 0 = 0
5 x 2 = 10	5 x 9 = 45	5 x 5 = 25	5 x 3 = 15	5 x 4 = 20	5 x 8 = 40
5 x 9 = 45	5 x 0 = 0	5 x 10 = 50	5 x 3 = 15	5 x 5 = 25	5 x 9 = 45
5 x 8 = 40	5 x 9 = 45	5 x 4 = 20	5 x 3 = 15	5 x 2 = 10	5 x 0 = 0
1 x 5 = 5	10 x 5 = 50	0 x 5 = 0	6 x 5 = 30	5 x 5 = 25	4 x 5 = 20
7 x 5 = 35	4 x 5 = 20	2 x 5 = 10	9 x 5 = 45	10 x 5 = 50	0 x 5 = 0
9 x 5 = 45	6 x 5 = 30	3 x 5 = 15	5 x 5 = 25	7 x 5 = 35	6 x 5 = 30
2 x 5 = 10	3 x 5 = 15	8 x 5 = 40	7 x 5 = 35	9 x 5 = 45	0 x 5 = 0
10 x 5 = 50	8 x 5 = 40	9 x 5 = 45	3 x 5 = 15	5 x 5 = 25	7 x 5 = 35

5 Facts Answer Keys

Page 46

I can multiply by 5s: Top AND Bottom

Name: _____ Date: _____
Goal: _____ problems in _____ seconds/minutes

5 × 5 = **25**	5 × 0 = **0**	5 × 6 = **30**	5 × 5 = **25**	8 × 5 = **40**	5 × 8 = **40**
5 × 9 = **45**	5 × 1 = **5**	2 × 5 = **10**	8 × 5 = **40**	5 × 7 = **35**	5 × 9 = **45**
10 × 5 = **50**	8 × 5 = **40**	5 × 9 = **45**	3 × 5 = **15**	5 × 7 = **35**	9 × 5 = **45**
1 × 5 = **5**	0 × 5 = **0**	3 × 5 = **15**	5 × 2 = **10**	5 × 5 = **25**	5 × 7 = **35**
5 × 6 = **30**	5 × 1 = **5**	5 × 0 = **0**	10 × 5 = **50**	5 × 8 = **40**	3 × 5 = **15**
2 × 5 = **10**	4 × 5 = **20**	5 × 4 = **20**	7 × 5 = **35**	5 × 6 = **30**	8 × 5 = **40**
5 × 2 = **10**	5 × 7 = **35**	0 × 5 = **0**	5 × 2 = **10**	7 × 5 = **35**	5 × 9 = **45**
5 × 8 = **40**	6 × 5 = **30**	3 × 5 = **15**	5 × 9 = **45**	5 × 4 = **20**	3 × 5 = **15**
1 × 5 = **5**	5 × 7 = **35**	5 × 9 = **45**	3 × 5 = **15**	2 × 5 = **10**	5 × 7 = **35**
5 × 5 = **25**	6 × 5 = **30**	5 × 9 = **45**	10 × 5 = **50**	5 × 3 = **15**	5 × 9 = **45**

Page 47

REVIEW: 0-5

Name: _____ Date: _____
Goal: _____ problems in _____ seconds/minutes

1 × 7 = **7**	5 × 3 = **15**	8 × 3 = **24**	2 × 7 = **14**	6 × 5 = **30**	0 × 8 = **0**
2 × 3 = **6**	5 × 9 = **45**	7 × 1 = **7**	0 × 4 = **0**	4 × 8 = **32**	5 × 5 = **25**
3 × 4 = **12**	4 × 7 = **28**	0 × 6 = **0**	6 × 3 = **18**	9 × 3 = **27**	10 × 5 = **50**
10 × 3 = **30**	0 × 7 = **0**	3 × 9 = **27**	5 × 2 = **10**	2 × 8 = **16**	0 × 3 = **0**
9 × 0 = **0**	4 × 4 = **16**	6 × 5 = **30**	5 × 9 = **45**	2 × 6 = **12**	1 × 5 = **5**
4 × 3 = **12**	8 × 5 = **40**	0 × 2 = **0**	1 × 10 = **10**	3 × 8 = **24**	5 × 8 = **40**
4 × 9 = **36**	7 × 2 = **14**	9 × 3 = **27**	5 × 10 = **50**	4 × 2 = **8**	8 × 3 = **24**
4 × 0 = **0**	10 × 2 = **20**	8 × 5 = **40**	3 × 7 = **21**	3 × 9 = **6**	0 × 1 = **0**
3 × 7 = **21**	9 × 2 = **18**	6 × 5 = **30**	4 × 4 = **16**	8 × 0 = **0**	2 × 7 = **14**
3 × 10 = **30**	0 × 2 = **0**	8 × 5 = **40**	5 × 8 = **40**	3 × 9 = **27**	2 × 2 = **4**

Page 48

REVIEW: 0-5

Name: _____ Date: _____
Goal: _____ problems in _____ seconds/minutes

3 × 4 = **12**	7 × 2 = **14**	0 × 7 = **0**	4 × 6 = **24**	9 × 1 = **9**	8 × 5 = **40**
7 × 4 = **28**	10 × 5 = **50**	6 × 3 = **18**	3 × 6 = **18**	8 × 4 = **32**	5 × 9 = **45**
2 × 1 = **2**	0 × 8 = **0**	4 × 5 = **20**	3 × 7 = **21**	8 × 2 = **16**	1 × 9 = **9**
3 × 10 = **30**	5 × 6 = **30**	9 × 4 = **36**	5 × 5 = **25**	6 × 2 = **12**	2 × 8 = **16**
0 × 6 = **0**	4 × 0 = **0**	3 × 3 = **9**	8 × 4 = **32**	1 × 9 = **9**	8 × 5 = **40**
3 × 3 = **9**	2 × 3 = **6**	6 × 4 = **24**	1 × 3 = **3**	10 × 3 = **30**	0 × 0 = **0**
5 × 4 = **20**	7 × 5 = **35**	3 × 9 = **27**	2 × 7 = **14**	0 × 1 = **0**	7 × 3 = **21**
0 × 7 = **0**	5 × 10 = **50**	6 × 4 = **24**	2 × 9 = **18**	3 × 8 = **24**	7 × 4 = **28**
3 × 3 = **9**	4 × 2 = **8**	5 × 2 = **10**	7 × 0 = **0**	0 × 0 = **0**	2 × 8 = **16**
10 × 4 = **40**	8 × 5 = **40**	3 × 9 = **27**	7 × 2 = **14**	5 × 4 = **20**	2 × 3 = **6**

Page 49

REVIEW: 0-5

Name: _____ Date: _____
Goal: _____ problems in _____ seconds/minutes

8 × 5 = **40**	4 × 3 = **12**	4 × 4 = **16**	8 × 2 = **16**	10 × 3 = **30**	7 × 0 = **0**
6 × 5 = **30**	4 × 8 = **32**	9 × 2 = **18**	1 × 0 = **0**	2 × 10 = **20**	7 × 2 = **14**
2 × 2 = **4**	5 × 2 = **10**	5 × 7 = **35**	2 × 9 = **18**	3 × 7 = **21**	5 × 5 = **25**
4 × 5 = **20**	4 × 10 = **40**	0 × 7 = **0**	6 × 4 = **24**	4 × 8 = **32**	7 × 3 = **21**
2 × 1 = **2**	4 × 1 = **4**	1 × 1 = **1**	0 × 1 = **0**	7 × 5 = **35**	4 × 8 = **32**
0 × 6 = **0**	4 × 6 = **24**	2 × 8 = **16**	7 × 1 = **7**	4 × 6 = **24**	2 × 6 = **12**
9 × 5 = **45**	2 × 0 = **0**	3 × 5 = **15**	6 × 3 = **24**	9 × 1 = **9**	4 × 8 = **32**
6 × 5 = **30**	4 × 8 = **32**	2 × 9 = **18**	0 × 1 = **0**	10 × 5 = **50**	4 × 3 = **12**
3 × 4 = **12**	4 × 7 = **28**	2 × 9 = **18**	8 × 5 = **40**	4 × 7 = **28**	10 × 4 = **40**
1 × 5 = **5**	6 × 5 = **30**	4 × 5 = **20**	7 × 3 = **21**	2 × 4 = **8**	5 × 9 = **45**

Page 50

Our New Fact Is 6

6 Facts
6 x 0 = 0
6 x 1 = 6
6 x 2 = 12
6 x 3 = 18
6 x 4 = 24
6 x 5 = 32
6 x 6 = 36
6 x 7 = 42
6 x 8 = 48
6 x 9 = 54
6 x 10 = 60

Draw squiggly lines from the facts to the correct answers!

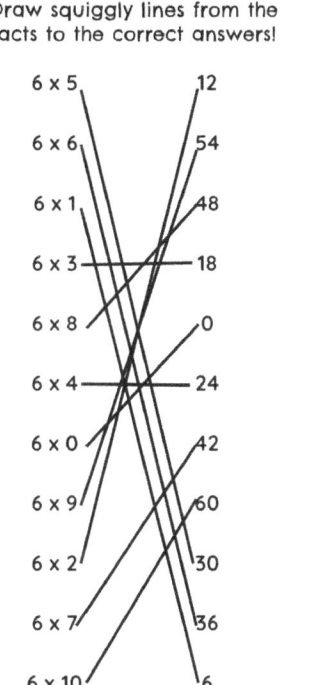

Page 51

Write Your 6 Facts

Trace it	Answer it	Fill in the blanks	Fill in the blanks	Write the fact
6 x 7 = 42	6 x 7 = 42	6 x 7 = 42	6 x 7 = 42	6 x 7 = 42
6 x 3 = 18	6 x 3 = 18	6 x 3 = 18	6 x 3 = 18	6 x 3 = 18
6 x 6 = 36	6 x 6 = 36	6 x 6 = 36	6 x 6 = 36	6 x 6 = 36
6 x 2 = 12	6 x 2 = 12	6 x 2 = 12	6 x 2 = 12	6 x 2 = 12
6 x 4 = 24	6 x 4 = 24	6 x 4 = 24	6 x 4 = 24	6 x 4 = 24
6 x 8 = 18	6 x 8 = 18	6 x 8 = 18	6 x 8 = 18	6 x 8 = 18
6 x 1 = 6	6 x 1 = 6	6 x 1 = 6	6 x 1 = 6	6 x 1 = 6
6 x 9 = 54	6 x 9 = 54	6 x 9 = 54	6 x 9 = 54	6 x 9 = 54
6 x 5 = 30	6 x 5 = 30	6 x 5 = 30	6 x 5 = 30	6 x 5 = 30
6 x 10 = 60	6 x 10 = 60	6 x 10 = 60	6 x 10 = 60	6 x 10 = 60
6 x 0 = 0	6 x 0 = 0	6 x 0 = 0	6 x 0 = 0	6 x 0 = 0

Page 52

Let's Practice!

Let's make sure you have all your facts down! Answer each multiplication problem.

$$\begin{array}{r} 6 \\ \times\ 6 \\ \hline 36 \end{array}$$

$$\begin{array}{r} 6 \\ \times\ 7 \\ \hline 42 \end{array}$$

$$\begin{array}{r} 6 \\ \times\ 8 \\ \hline 48 \end{array}$$

$$\begin{array}{r} 6 \\ \times\ 5 \\ \hline 30 \end{array}$$

$$\begin{array}{r} 6 \\ \times\ 9 \\ \hline 54 \end{array}$$

$$\begin{array}{r} 6 \\ \times\ 0 \\ \hline 0 \end{array}$$

$$\begin{array}{r} 6 \\ \times\ 4 \\ \hline 24 \end{array}$$

$$\begin{array}{r} 6 \\ \times\ 1 \\ \hline 6 \end{array}$$

$$\begin{array}{r} 6 \\ \times\ 3 \\ \hline 18 \end{array}$$

$$\begin{array}{r} 6 \\ \times\ 2 \\ \hline 12 \end{array}$$

$$\begin{array}{r} 6 \\ \times\ 10 \\ \hline 60 \end{array}$$

Page 53

I can multiply by 6: TOP or BOTTOM	Name: _____ Date: _____ Goal: ___ problems in ___ seconds/minutes

6 x 5	6 x 7	6 x 0	6 x 10	6 x 3	6 x 2
30	42	0	60	18	12
6 x 8	6 x 6	6 x 4	6 x 6	6 x 5	6 x 8
48	36	24	36	30	48
6 x 9	6 x 10	6 x 0	6 x 1	6 x 2	6 x 5
54	50	0	6	12	30
6 x 9	6 x 8	6 x 5	6 x 3	6 x 7	6 x 8
36	48	30	18	42	48
6 x 10	6 x 2	6 x 9	6 x 1	6 x 7	6 x 5
60	12	54	6	42	30
3 x 6	8 x 6	7 x 6	5 x 6	4 x 6	10 x 6
18	48	42	30	24	60
7 x 6	3 x 6	2 x 6	3 x 6	8 x 6	7 x 6
42	18	12	18	48	42
10 x 6	0 x 6	2 x 6	1 x 6	8 x 6	9 x 6
60	0	12	6	48	54
6 x 6	3 x 6	6 x 6	9 x 6	7 x 6	1 x 6
36	18	36	54	42	6
10 x 6	0 x 6	4 x 6	9 x 6	2 x 6	4 x 6
60	0	24	54	12	24

6 Facts Answer Keys

Page 54

I can multiply by 6s: Top AND Bottom

Name: _____ Date: _____
Goal: _____ problems in _____ seconds/minutes

10 × 6 = **60**	6 × 0 = **0**	1 × 6 = **6**	6 × 8 = **48**	9 × 6 = **54**	6 × 5 = **30**
6 × 2 = **12**	6 × 1 = **6**	7 × 6 = **42**	6 × 9 = **54**	10 × 6 = **60**	2 × 6 = **12**
6 × 9 = **54**	7 × 6 = **42**	4 × 6 = **24**	6 × 6 = **36**	6 × 1 = **6**	5 × 6 = **30**
9 × 6 = **54**	7 × 6 = **42**	6 × 3 = **18**	5 × 6 = **30**	6 × 8 = **48**	6 × 2 = **12**
6 × 8 = **48**	6 × 0 = **0**	10 × 6 = **60**	6 × 8 = **48**	9 × 6 = **54**	6 × 6 = **36**
2 × 6 = **12**	6 × 8 = **48**	6 × 2 = **12**	6 × 4 = **24**	6 × 7 = **42**	9 × 6 = **54**
6 × 10 = **60**	8 × 6 = **48**	6 × 3 = **18**	6 × 5 = **30**	6 × 3 = **18**	8 × 6 = **48**
9 × 6 = **54**	4 × 6 = **24**	6 × 3 = **18**	7 × 6 = **42**	9 × 6 = **54**	1 × 6 = **6**
6 × 0 = **0**	6 × 10 = **60**	6 × 6 = **36**	6 × 4 = **24**	7 × 6 = **42**	6 × 2 = **12**
8 × 6 = **48**	6 × 7 = **42**	6 × 5 = **30**	9 × 6 = **54**	6 × 6 = **36**	1 × 6 = **6**

Page 55

REVIEW: 0-6

Name: _____ Date: _____
Goal: _____ problems in _____ seconds/minutes

7 × 6 = **42**	3 × 6 = **18**	8 × 1 = **8**	9 × 0 = **0**	2 × 6 = **12**	6 × 4 = **24**
5 × 3 = **15**	8 × 2 = **16**	10 × 6 = **60**	2 × 3 = **6**	7 × 3 = **21**	5 × 9 = **45**
2 × 10 = **20**	3 × 0 = **0**	5 × 8 = **40**	6 × 6 = **36**	9 × 6 = **54**	3 × 1 = **3**
8 × 3 = **24**	4 × 6 = **24**	0 × 1 = **0**	5 × 8 = **40**	3 × 0 = **0**	6 × 10 = **60**
8 × 5 = **40**	3 × 7 = **21**	2 × 9 = **18**	8 × 4 = **24**	6 × 5 = **30**	0 × 9 = **0**
1 × 6 = **6**	8 × 0 = **0**	5 × 7 = **35**	6 × 7 = **42**	9 × 2 = **18**	8 × 3 = **24**
4 × 4 = **16**	3 × 7 = **21**	3 × 9 = **27**	2 × 0 = **0**	9 × 4 = **36**	5 × 5 = **25**
2 × 1 = **2**	1 × 7 = **7**	3 × 9 = **27**	0 × 3 = **0**	5 × 6 = **30**	8 × 3 = **24**
9 × 2 = **18**	4 × 7 = **28**	6 × 7 = **42**	9 × 3 = **27**	6 × 8 = **48**	2 × 2 = **4**
5 × 2 = **10**	4 × 5 = **20**	6 × 4 = **24**	4 × 6 = **24**	9 × 6 = **54**	3 × 3 = **9**

Page 56

REVIEW: 0-6

Name: _____ Date: _____
Goal: _____ problems in _____ seconds/minutes

5 × 5 = **25**	3 × 6 = **18**	9 × 2 = **18**	0 × 4 = **0**	5 × 9 = **45**	8 × 4 = **32**
1 × 3 = **3**	3 × 5 = **15**	2 × 2 = **4**	7 × 0 = **0**	9 × 6 = **54**	8 × 5 = **40**
0 × 10 = **0**	7 × 6 = **42**	5 × 9 = **45**	3 × 6 = **18**	6 × 3 = **18**	2 × 6 = **12**
9 × 4 = **36**	4 × 4 = **16**	6 × 6 = **36**	2 × 9 = **18**	0 × 8 = **0**	5 × 3 = **15**
1 × 7 = **7**	7 × 5 = **35**	6 × 4 = **24**	9 × 3 = **27**	2 × 7 = **14**	6 × 3 = **18**
7 × 0 = **0**	3 × 10 = **30**	9 × 0 = **0**	2 × 6 = **12**	7 × 3 = **21**	9 × 6 = **54**
6 × 7 = **42**	9 × 3 = **27**	7 × 2 = **14**	6 × 5 = **30**	5 × 5 = **25**	9 × 2 = **18**
2 × 6 = **12**	5 × 7 = **35**	3 × 0 = **0**	10 × 6 = **60**	4 × 5 = **20**	3 × 8 = **24**
2 × 2 = **4**	6 × 2 = **12**	8 × 3 = **24**	9 × 5 = **45**	5 × 8 = **40**	3 × 7 = **21**
2 × 8 = **16**	10 × 5 = **50**	4 × 6 = **24**	7 × 2 = **14**	4 × 9 = **36**	6 × 6 = **36**

Page 57

REVIEW: 0-6

Name: _____ Date: _____
Goal: _____ problems in _____ seconds/minutes

9 × 4 = **36**	0 × 5 = **0**	2 × 9 = **18**	7 × 3 = **21**	4 × 6 = **24**	2 × 2 = **4**
7 × 6 = **42**	4 × 8 = **32**	9 × 2 = **18**	1 × 0 = **0**	1 × 1 = **1**	8 × 3 = **24**
2 × 8 = **16**	3 × 3 = **9**	5 × 8 = **40**	6 × 7 = **42**	5 × 5 = **25**	9 × 3 = **27**
3 × 2 = **6**	2 × 2 = **4**	7 × 4 = **28**	1 × 5 = **5**	9 × 0 = **0**	3 × 7 = **21**
8 × 4 = **32**	3 × 3 = **9**	5 × 5 = **25**	2 × 6 = **12**	9 × 4 = **36**	5 × 3 = **15**
7 × 4 = **28**	4 × 4 = **16**	3 × 7 = **21**	9 × 2 = **18**	5 × 8 = **40**	10 × 5 = **50**
6 × 6 = **36**	7 × 5 = **35**	4 × 4 = **16**	9 × 3 = **27**	0 × 6 = **0**	10 × 4 = **40**
2 × 3 = **6**	3 × 3 = **9**	5 × 4 = **20**	6 × 7 = **42**	9 × 3 = **27**	2 × 7 = **14**
3 × 8 = **24**	5 × 5 = **25**	3 × 5 = **15**	6 × 2 = **12**	9 × 0 = **0**	5 × 10 = **50**
3 × 7 = **21**	9 × 2 = **17**	4 × 2 = **8**	5 × 0 = **0**	7 × 3 = **21**	5 × 4 = **20**

7 Facts Answer Keys

Page 58

Our New Fact Is

7 Facts
7 x 0 = 0
7 x 1 = 7
7 x 2 = 14
7 x 3 = 21
7 x 4 = 28
7 x 5 = 35
7 x 6 = 42
7 x 7 = 49
7 x 8 = 56
7 x 9 = 63
7 x 10 = 70

Draw straight lines from the facts to the correct answers!

7 x 2	70
7 x 7	21
7 x 6	28
7 x 0	63
7 x 5	14
7 x 1	35
7 x 10	56
7 x 4	0
7 x 8	7
7 x 9	42
7 x 3	49

Page 59

Write Your 7 Facts

Trace it	Answer it	Fill in the blanks	Fill in the blanks	Write the fact
7 x 8 = 56	7 x 8 = 56	7 x 8 = 56	7 x 8 = 56	7 x 8 = 56
7 x 3 = 21	7 x 3 = 21	7 x 3 = 21	7 x 3 = 21	7 x 3 = 21
7 x 10 = 70	7 x 10 = 70	7 x 10 = 70	7 x 10 = 70	7 x 10 = 70
7 x 2 = 14	7 x 2 = 14	7 x 2 = 14	7 x 2 = 14	7 x 2 = 14
7 x 9 = 63	7 x 9 = 63	7 x 9 = 63	7 x 9 = 63	7 x 9 = 63
7 x 7 = 49	7 x 7 = 49	7 x 7 = 49	7 x 7 = 49	7 x 7 = 49
7 x 1 = 7	7 x 1 = 7	7 x 1 = 7	7 x 1 = 7	7 x 1 = 7
7 x 6 = 42	7 x 6 = 42	7 x 6 = 42	7 x 6 = 42	7 x 6 = 42
7 x 5 = 35	7 x 5 = 35	7 x 5 = 35	7 x 5 = 35	7 x 5 = 35
7 x 0 = 0	7 x 0 = 0	7 x 0 = 0	7 x 0 = 0	7 x 0 = 0
7 x 4 = 28	7 x 4 = 28	7 x 4 = 28	7 x 4 = 28	7 x 4 = 28

Page 60

Let's Practice!

Let's make sure you have all your facts down! Answer each multiplication problem.

$$7 \times 4 = 28$$
$$7 \times 9 = 63$$
$$7 \times 5 = 35$$
$$7 \times 6 = 42$$
$$7 \times 1 = 7$$
$$7 \times 7 = 49$$
$$7 \times 8 = 56$$
$$7 \times 0 = 0$$
$$7 \times 2 = 14$$
$$7 \times 10 = 70$$
$$7 \times 3 = 21$$

Page 61

I can multiply by 7: TOP or BOTTOM

Name: _____ Date: _____
Goal: ___ problems in ___ seconds/minutes

7 × 9 = 63	7 × 6 = 42	7 × 2 = 14	7 × 7 = 49	7 × 0 = 0	7 × 10 = 14
7 × 2 = 14	7 × 8 = 56	7 × 6 = 42	7 × 4 = 28	7 × 5 = 35	7 × 8 = 56
7 × 1 = 7	7 × 8 = 56	7 × 7 = 49	7 × 7 = 49	7 × 9 = 63	7 × 1 = 7
7 × 3 = 21	7 × 0 = 0	7 × 10 = 70	7 × 2 = 14	7 × 7 = 49	7 × 9 = 63
7 × 5 = 35	7 × 4 = 28	7 × 6 = 42	7 × 3 = 21	7 × 1 = 7	7 × 0 = 0
10 × 7 = 70	2 × 7 = 14	8 × 7 = 56	4 × 7 = 28	3 × 7 = 21	8 × 7 = 56
5 × 7 = 35	8 × 7 = 56	9 × 7 = 63	4 × 7 = 28	5 × 7 = 35	1 × 7 = 7
0 × 7 = 0	8 × 7 = 56	3 × 7 = 21	10 × 7 = 70	4 × 7 = 28	8 × 7 = 56
8 × 7 = 56	3 × 7 = 21	5 × 7 = 35	6 × 7 = 42	7 × 7 = 49	2 × 7 = 14
1 × 7 = 7	9 × 7 = 63	7 × 7 = 49	6 × 7 = 42	8 × 7 = 56	0 × 7 = 0

7 Facts Answer Keys

Page 62

I can multiply by 7s: Top AND Bottom

Name: _____ Date: _____
Goal: _____ problems in _____ seconds/minutes

7 × 8 = 56	4 × 7 = 28	2 × 7 = 14	7 × 8 = 56	0 × 7 = 0	7 × 10 = 70
2 × 7 = 14	5 × 7 = 35	7 × 8 = 56	7 × 6 = 42	2 × 7 = 14	7 × 9 = 63
4 × 7 = 28	7 × 5 = 35	7 × 0 = 0	7 × 10 = 70	9 × 7 = 63	4 × 7 = 28
7 × 2 = 14	7 × 8 = 56	9 × 7 = 63	2 × 7 = 14	7 × 7 = 49	7 × 4 = 28
7 × 3 = 21	3 × 7 = 21	7 × 6 = 42	9 × 7 = 63	7 × 1 = 7	10 × 7 = 70
2 × 7 = 14	9 × 7 = 63	7 × 7 = 49	4 × 7 = 28	3 × 7 = 21	7 × 2 = 14
7 × 7 = 49	7 × 2 = 14	7 × 4 = 28	7 × 3 = 21	10 × 7 = 70	0 × 7 = 0
7 × 1 = 7	9 × 7 = 63	8 × 7 = 56	3 × 7 = 21	7 × 2 = 14	5 × 7 = 35
4 × 7 = 28	7 × 9 = 63	7 × 7 = 49	7 × 6 = 42	5 × 7 = 35	7 × 3 = 21
7 × 2 = 14	0 × 7 = 0	2 × 7 = 14	7 × 1 = 7	8 × 7 = 56	7 × 4 = 28

Page 63

REVIEW: 0-7

Name: _____ Date: _____
Goal: _____ problems in _____ seconds/minutes

0 × 0 = 0	8 × 6 = 48	7 × 3 = 21	9 × 7 = 63	6 × 3 = 18	2 × 0 = 0
10 × 7 = 70	3 × 6 = 18	5 × 7 = 35	9 × 2 = 18	4 × 7 = 28	3 × 8 = 24
7 × 7 = 49	9 × 7 = 63	7 × 2 = 14	4 × 7 = 28	9 × 4 = 36	3 × 10 = 30
4 × 9 = 36	0 × 4 = 0	9 × 7 = 63	10 × 7 = 70	5 × 3 = 15	2 × 7 = 14
4 × 7 = 28	2 × 8 = 16	0 × 3 = 0	4 × 10 = 40	8 × 7 = 56	4 × 5 = 20
3 × 8 = 24	9 × 7 = 63	1 × 0 = 0	9 × 6 = 54	3 × 7 = 21	3 × 3 = 9
2 × 9 = 18	0 × 7 = 0	4 × 7 = 28	6 × 2 = 12	5 × 6 = 30	0 × 5 = 0
7 × 5 = 35	9 × 7 = 63	1 × 6 = 6	3 × 9 = 27	0 × 4 = 0	7 × 5 = 35
6 × 7 = 42	8 × 7 = 56	2 × 6 = 12	0 × 5 = 0	3 × 6 = 18	8 × 4 = 32
3 × 9 = 27	10 × 7 = 70	5 × 4 = 20	8 × 3 = 24	4 × 5 = 20	2 × 9 = 18

Page 64

REVIEW: 0-7

Name: _____ Date: _____
Goal: _____ problems in _____ seconds/minutes

6 × 6 = 36	9 × 7 = 63	3 × 5 = 15	6 × 6 = 36	0 × 8 = 0	7 × 1 = 7
2 × 5 = 10	8 × 7 = 56	5 × 5 = 25	6 × 4 = 24	7 × 9 = 63	2 × 1 = 2
4 × 5 = 20	4 × 4 = 16	7 × 2 = 14	7 × 7 = 49	0 × 1 = 0	8 × 4 = 32
2 × 0 = 0	5 × 4 = 20	3 × 4 = 12	6 × 9 = 54	0 × 3 = 0	10 × 3 = 30
4 × 4 = 16	7 × 6 = 42	0 × 2 = 0	7 × 6 = 42	4 × 8 = 32	4 × 8 = 32
2 × 6 = 12	4 × 7 = 28	8 × 7 = 56	10 × 4 = 40	3 × 5 = 15	6 × 2 = 12
3 × 6 = 18	8 × 3 = 24	7 × 5 = 35	5 × 9 = 45	2 × 10 = 20	8 × 4 = 32
3 × 6 = 18	7 × 5 = 35	6 × 3 = 15	2 × 9 = 18	6 × 6 = 36	4 × 0 = 0
3 × 4 = 12	5 × 2 = 10	8 × 1 = 8	1 × 7 = 7	6 × 3 = 18	5 × 8 = 40
3 × 3 = 9	6 × 7 = 42	4 × 9 = 36	3 × 7 = 21	6 × 5 = 30	2 × 7 = 14

Page 65

REVIEW: 0-7

Name: _____ Date: _____
Goal: _____ problems in _____ seconds/minutes

5 × 5 = 25	0 × 3 = 0	10 × 7 = 70	4 × 6 = 24	5 × 2 = 10	9 × 5 = 45
3 × 6 = 18	9 × 4 = 36	7 × 9 = 63	2 × 7 = 14	6 × 3 = 18	1 × 1 = 1
6 × 9 = 54	3 × 7 = 21	0 × 3 = 0	10 × 5 = 50	4 × 4 = 16	3 × 7 = 21
9 × 2 = 18	5 × 7 = 35	3 × 9 = 27	10 × 4 = 40	2 × 8 = 16	6 × 1 = 6
4 × 4 = 16	8 × 3 = 24	6 × 9 = 54	0 × 1 = 0	5 × 6 = 30	3 × 9 = 27
2 × 7 = 14	6 × 6 = 36	4 × 9 = 36	3 × 10 = 30	5 × 3 = 15	8 × 7 = 56
3 × 3 = 9	7 × 5 = 35	4 × 4 = 16	9 × 2 = 18	4 × 7 = 28	7 × 7 = 49
2 × 6 = 18	5 × 3 = 15	4 × 9 = 36	3 × 9 = 27	5 × 5 = 25	0 × 3 = 0
2 × 1 = 2	7 × 5 = 35	9 × 3 = 27	5 × 7 = 35	1 × 9 = 9	0 × 9 = 0
8 × 5 = 45	2 × 7 = 14	10 × 6 = 60	3 × 8 = 24	6 × 4 = 24	2 × 8 = 16

8 Facts Answer Keys

Page 66

Our New Fact Is 8

8 Facts
8 x 0 = 0
8 x 1 = 8
8 x 2 = 16
8 x 3 = 24
8 x 4 = 32
8 x 5 = 40
8 x 6 = 48
8 x 7 = 56
8 x 8 = 64
8 x 9 = 72
8 x 10 = 80

Draw dashed lines from the facts to the correct answers!

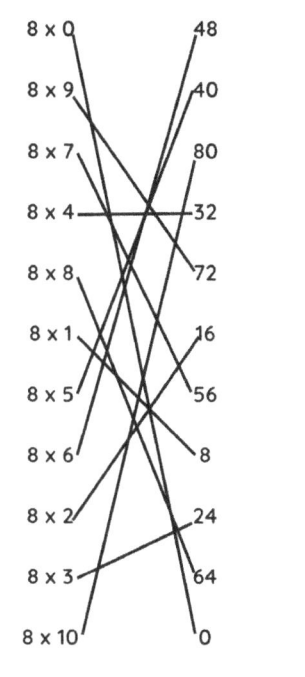

Page 67

Write Your 8 Facts

Trace it	Answer it	Fill in the blanks	Fill in the blanks	Write the fact
8 x 4 = 32	8 x 4 = 32	8 x 4 = 32	8 x 4 = 32	8 x 4 = 32
8 x 0 = 0	8 x 0 = 0	8 x 0 = 0	8 x 0 = 0	8 x 0 = 0
8 x 10 = 80	8 x 10 = 80	8 x 10 = 80	8 x 10 = 80	8 x 10 = 80
8 x 5 = 40	8 x 5 = 40	8 x 5 = 40	8 x 5 = 40	8 x 5 = 40
8 x 9 = 72	8 x 9 = 72	8 x 9 = 72	8 x 9 = 72	8 x 9 = 72
8 x 1 = 8	8 x 1 = 8	8 x 1 = 8	8 x 1 = 8	8 x 1 = 8
8 x 6 = 48	8 x 6 = 48	8 x 6 = 48	8 x 6 = 48	8 x 6 = 48
8 x 2 = 16	8 x 2 = 16	8 x 2 = 16	8 x 2 = 16	8 x 2 = 16
8 x 8 = 64	8 x 8 = 64	8 x 8 = 64	8 x 8 = 64	8 x 8 = 64
8 x 3 = 24	8 x 3 = 24	8 x 3 = 24	8 x 3 = 24	8 x 3 = 24
8 x 7 = 56	8 x 7 = 56	8 x 7 = 56	8 x 7 = 56	8 x 7 = 56

Page 68

Let's Practice!

Let's make sure you have all your facts down! Answer each multiplication problem.

$$8 \times 8 = 64$$

$$8 \times 7 = 56$$

$$8 \times 9 = 72$$

$$8 \times 3 = 24$$

$$8 \times 4 = 32$$

$$8 \times 10 = 80$$

$$8 \times 1 = 8$$

$$8 \times 2 = 16$$

$$8 \times 6 = 48$$

$$8 \times 0 = 80$$

$$8 \times 5 = 40$$

Page 69

I can multiply by 8: TOP or BOTTOM

Name: _____   Date: _____
Goal: _____ problems in _____ seconds/minutes

8 x 0 = 0	8 x 2 = 16	8 x 1 = 8	8 x 10 = 80	8 x 8 = 64	8 x 6 = 48
8 x 6 = 48	8 x 3 = 24	8 x 2 = 16	8 x 8 = 64	8 x 9 = 72	8 x 4 = 32
8 x 5 = 40	8 x 7 = 56	8 x 6 = 48	8 x 1 = 8	8 x 9 = 72	8 x 1 = 8
8 x 3 = 24	8 x 2 = 16	8 x 3 = 24	8 x 5 = 40	8 x 8 = 64	8 x 9 = 72
8 x 7 = 56	8 x 3 = 24	8 x 8 = 64	8 x 0 = 0	8 x 1 = 8	8 x 10 = 80
7 x 8 = 56	3 x 8 = 24	4 x 8 = 32	5 x 8 = 40	2 x 8 = 16	10 x 8 = 80
0 x 8 = 0	3 x 8 = 24	5 x 8 = 40	7 x 8 = 56	9 x 8 = 72	4 x 8 = 32
2 x 8 = 16	1 x 8 = 8	10 x 8 = 80	4 x 8 = 32	3 x 8 = 24	2 x 8 = 16
8 x 8 = 64	7 x 8 = 56	6 x 8 = 48	5 x 8 = 40	9 x 8 = 72	0 x 8 = 0
10 x 8 = 80	2 x 8 = 16	8 x 8 = 64	3 x 8 = 24	5 x 8 = 40	2 x 8 = 16

Page 70

I can multiply by 8s: Top AND Bottom

Name: _____ Date: _____
Goal: _____ problems in _____ seconds/minutes

64	56	16	64	72	0
8	16	48	40	72	40
72	24	56	72	8	80
0	64	56	24	16	8
64	56	24	40	72	0
80	16	32	40	24	48
16	48	72	48	40	8
0	64	64	0	80	16
24	56	48	24	32	24
56	40	72	24	32	40

Page 71

REVIEW: 0-8

Name: _____ Date: _____
Goal: _____ problems in _____ seconds/minutes

32	72	18	16	0	64
21	24	40	0	80	12
25	48	0	7	72	6
70	12	10	9	35	9
16	24	0	32	42	27
21	32	18	0	50	6
72	24	35	18	28	5
32	35	28	18	30	12
15	12	63	70	24	12
9	30	20	24	5	72

Page 72

REVIEW: 0-8

Name: _____ Date: _____
Goal: _____ problems in _____ seconds/minutes

28	48	20	0	50	20
54	0	4	12	56	30
15	14	56	40	2	63
80	35	21	9	0	14
64	20	18	35	24	50
10	24	27	80	35	8
15	16	63	24	40	8
35	24	36	10	16	14
27	56	35	0	14	32
2	15	12	32	8	25

Page 73

REVIEW: 0-8

Name: _____ Date: _____
Goal: _____ problems in _____ seconds/minutes

18	35	24	12	70	24
12	35	15	18	48	0
10	30	27	56	72	4
1	15	42	15	0	40
30	32	12	45	0	30
64	35	12	12	45	15
24	49	15	63	12	4
12	10	35	24	5	12
48	24	18	0	60	9
35	18	30	0	30	24

9 Facts Answer Keys

Page 74

Our New Fact Is 9

9 Facts	
9 x 0	= 0
9 x 1	= 9
9 x 2	= 18
9 x 3	= 27
9 x 4	= 37
9 x 5	= 45
9 x 6	= 54
9 x 7	= 63
9 x 8	= 72
9 x 9	= 81
9 x 10	= 90

Draw squiggly lines from the facts to the correct answers!

9 x 5 — 0
9 x 9 — 9
9 x 6 — 81
9 x 0 — 18
9 x 1 — 72
9 x 7 — 27
9 x 2 — 36
9 x 10 — 90
9 x 3 — 63
9 x 4 — 45
9 x 8 — 54

Page 75

Write Your 9 Facts

Trace it	Answer it	Fill in the blanks	Fill in the blanks	Write the fact
9 x 5 = 45	9 x 5 = 45	9 x 5 = 45	9 x 5 = 45	9 x 5 = 45
9 x 6 = 54	9 x 6 = 54	9 x 6 = 54	9 x 6 = 54	9 x 6 = 54
9 x 1 = 9	9 x 1 = 9	9 x 1 = 9	9 x 1 = 9	9 x 1 = 9
9 x 4 = 36	9 x 4 = 36	9 x 4 = 36	9 x 4 = 36	9 x 4 = 36
9 x 0 = 0	9 x 0 = 0	9 x 0 = 0	9 x 0 = 0	9 x 0 = 0
9 x 7 = 63	9 x 7 = 63	9 x 7 = 63	9 x 7 = 63	9 x 7 = 63
9 x 3 = 27	9 x 3 = 27	9 x 3 = 27	9 x 3 = 27	9 x 3 = 27
9 x 2 = 18	9 x 2 = 18	9 x 2 = 18	9 x 2 = 18	9 x 2 = 18
9 x 8 = 72	9 x 8 = 72	9 x 8 = 72	9 x 8 = 72	9 x 8 = 72
9 x 10 = 90	9 x 10 = 90	9 x 10 = 90	9 x 10 = 90	9 x 10 = 90
9 x 9 = 81	9 x 9 = 81	9 x 9 = 81	9 x 9 = 81	9 x 9 = 81

Page 76

Let's Practice!

Let's make sure you have all your facts down! Answer each multiplication problem

$$9 \times 10 = 90$$

$$9 \times 3 = 27$$

$$9 \times 4 = 36$$

$$9 \times 6 = 54$$

$$9 \times 7 = 63$$

$$9 \times 9 = 81$$

$$9 \times 8 = 72$$

$$9 \times 5 = 45$$

$$9 \times 1 = 9$$

$$9 \times 2 = 18$$

$$9 \times 0 = 0$$

Page 77

I can multiply by 9: TOP or BOTTOM

Name: _____ Date: _____

Goal: _____ problems in _____ seconds/minutes

9 x 0 = 0	9 x 8 = 72	9 x 2 = 18	9 x 4 = 36	9 x 6 = 54	9 x 3 = 27
9 x 7 = 63	9 x 4 = 36	9 x 3 = 27	9 x 7 = 63	9 x 0 = 0	9 x 10 = 90
9 x 3 = 27	9 x 8 = 72	9 x 5 = 45	9 x 2 = 18	9 x 9 = 81	9 x 10 = 90
9 x 8 = 72	9 x 5 = 45	9 x 0 = 0	9 x 2 = 18	9 x 5 = 45	9 x 7 = 63
9 x 7 = 63	9 x 3 = 27	9 x 5 = 45	9 x 8 = 72	9 x 7 = 63	9 x 10 = 90
10 x 9 = 90	2 x 9 = 18	8 x 9 = 72	9 x 9 = 81	5 x 9 = 45	6 x 9 = 54
3 x 9 = 27	4 x 9 = 32	9 x 9 = 81	7 x 9 = 63	8 x 9 = 72	0 x 9 = 0
2 x 9 = 18	10 x 9 = 90	3 x 9 = 27	5 x 9 = 45	8 x 9 = 72	6 x 9 = 54
4 x 9 = 36	2 x 9 = 18	4 x 9 = 36	1 x 9 = 9	0 x 9 = 0	6 x 9 = 54
2 x 9 = 18	7 x 9 = 63	10 x 9 = 90	0 x 9 = 0	4 x 9 = 36	3 x 9 = 27

9 Facts Answer Keys

Page 78

I can multiply by 9s: Top AND Bottom

Name: ____________ Date: ________

Goal: ____ problems in ____ seconds/minutes

3 × 9 = 27	9 × 2 = 18	9 × 9 = 81	9 × 0 = 0	10 × 9 = 90	9 × 3 = 27
9 × 5 = 45	9 × 3 = 27	8 × 9 = 72	9 × 5 = 45	6 × 9 = 54	8 × 9 = 72
9 × 10 = 90	8 × 9 = 72	0 × 9 = 0	1 × 9 = 9	9 × 2 = 18	4 × 9 = 36
3 × 9 = 27	7 × 9 = 63	9 × 8 = 72	4 × 9 = 36	9 × 8 = 72	9 × 0 = 0
9 × 1 = 9	9 × 2 = 18	9 × 9 = 81	9 × 6 = 54	5 × 9 = 45	4 × 9 = 36
3 × 9 = 27	9 × 6 = 54	9 × 9 = 81	7 × 9 = 63	9 × 4 = 36	3 × 9 = 27
9 × 10 = 90	9 × 9 = 81	9 × 3 = 27	9 × 2 = 18	9 × 5 = 45	10 × 9 = 90
4 × 9 = 9	7 × 9 = 63	9 × 8 = 72	8 × 9 = 72	9 × 9 = 81	0 × 9 = 0
9 × 2 = 18	9 × 6 = 54	1 × 9 = 9	9 × 10 = 90	7 × 9 = 63	9 × 4 = 36
3 × 9 = 27	9 × 2 = 18	9 × 8 = 72	6 × 9 = 54	4 × 9 = 36	9 × 7 = 63

Page 79

REVIEW: 0-9

Name: ____________ Date: ________

Goal: ____ problems in ____ seconds/minutes

7 × 9 = 63	4 × 6 = 24	9 × 9 = 81	1 × 5 = 5	4 × 7 = 28	7 × 0 = 0
4 × 7 = 28	9 × 1 = 9	3 × 5 = 15	6 × 2 = 12	9 × 8 = 72	5 × 5 = 25
2 × 5 = 10	7 × 7 = 49	0 × 4 = 0	9 × 6 = 54	3 × 7 = 21	1 × 6 = 6
9 × 9 = 81	5 × 3 = 15	7 × 5 = 35	6 × 9 = 54	2 × 0 = 0	3 × 6 = 18
4 × 9 = 36	5 × 9 = 45	0 × 2 = 0	9 × 10 = 90	5 × 3 = 15	2 × 6 = 12
9 × 4 = 36	7 × 2 = 14	0 × 7 = 0	5 × 3 = 15	4 × 8 = 32	10 × 5 = 50
3 × 6 = 18	7 × 2 = 14	9 × 5 = 45	7 × 8 = 56	3 × 9 = 27	2 × 0 = 0
0 × 5 = 0	4 × 6 = 24	8 × 9 = 72	9 × 10 = 90	4 × 2 = 8	5 × 4 = 20
8 × 5 = 40	9 × 3 = 27	1 × 8 = 8	6 × 4 = 24	6 × 6 = 36	6 × 2 = 12
0 × 9 = 0	4 × 2 = 8	1 × 6 = 6	8 × 7 = 56	4 × 9 = 36	5 × 6 = 30

Page 80

REVIEW: 0-9

Name: ____________ Date: ________

Goal: ____ problems in ____ seconds/minutes

3 × 5 = 15	6 × 7 = 42	2 × 8 = 16	0 × 2 = 0	7 × 10 = 70	6 × 8 = 48
9 × 2 = 18	6 × 5 = 30	3 × 7 = 21	0 × 5 = 0	6 × 2 = 12	1 × 8 = 8
6 × 6 = 36	8 × 3 = 24	7 × 7 = 49	5 × 2 = 10	3 × 9 = 27	0 × 0 = 0
2 × 6 = 12	5 × 4 = 20	7 × 8 = 56	2 × 6 = 12	9 × 5 = 45	2 × 4 = 8
8 × 4 = 32	5 × 6 = 30	3 × 8 = 24	9 × 9 = 81	4 × 7 = 28	6 × 2 = 12
8 × 7 = 56	4 × 6 = 24	5 × 7 = 35	9 × 4 = 36	2 × 6 = 12	4 × 4 = 16
3 × 7 = 21	8 × 4 = 32	3 × 6 = 18	5 × 5 = 25	9 × 8 = 72	2 × 5 = 10
5 × 6 = 30	4 × 4 = 16	8 × 2 = 16	0 × 8 = 0	5 × 0 = 0	4 × 7 = 28
9 × 2 = 18	6 × 4 = 24	7 × 3 = 21	2 × 5 = 10	9 × 7 = 63	6 × 7 = 42
2 × 4 = 8	5 × 2 = 10	8 × 4 = 32	6 × 5 = 30	3 × 9 = 27	3 × 10 = 30

Page 81

REVIEW: 0-9

Name: ____________ Date: ________

Goal: ____ problems in ____ seconds/minutes

8 × 5 = 40	3 × 2 = 6	6 × 9 = 54	4 × 0 = 0	7 × 3 = 21	4 × 4 = 16
3 × 5 = 15	2 × 2 = 4	6 × 7 = 42	8 × 3 = 24	9 × 5 = 45	9 × 9 = 81
10 × 8 = 80	5 × 6 = 30	4 × 7 = 28	8 × 8 = 64	3 × 2 = 6	6 × 5 = 30
5 × 6 = 30	3 × 3 = 9	7 × 8 = 56	4 × 9 = 36	8 × 2 = 16	1 × 3 = 3
7 × 7 = 49	4 × 3 = 12	5 × 6 = 30	9 × 4 = 36	10 × 2 = 20	5 × 4 = 20
4 × 6 = 24	5 × 5 = 35	7 × 2 = 14	1 × 9 = 9	5 × 10 = 50	6 × 4 = 24
2 × 3 = 6	5 × 3 = 15	6 × 2 = 12	5 × 4 = 20	8 × 8 = 64	7 × 6 = 42
5 × 6 = 30	2 × 7 = 14	8 × 7 = 56	6 × 5 = 30	4 × 7 = 28	9 × 9 = 81
3 × 5 = 15	4 × 2 = 8	1 × 2 = 2	3 × 8 = 24	7 × 6 = 42	4 × 9 = 36
0 × 5 = 0	3 × 4 = 12	2 × 7 = 14	8 × 6 = 42	5 × 3 = 15	6 × 5 = 30

Page 82

Our New Fact Is 10

10 Facts
10 x 0 = 0
10 x 1 = 10
10 x 2 = 20
10 x 3 = 30
10 x 4 = 40
10 x 5 = 50
10 x 6 = 60
10 x 7 = 70
10 x 8 = 80
10 x 9 = 90
10 x 10 = 100

Draw straight lines from the facts to the correct answers!

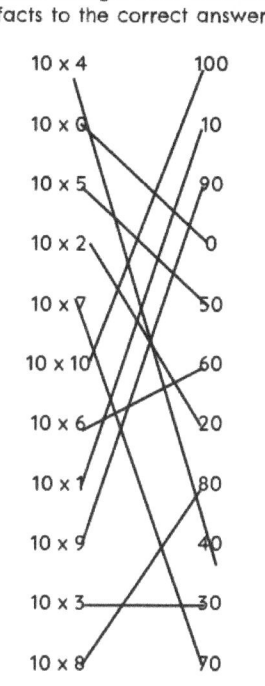

Page 83

Write Your 10 Facts

Trace it	Answer it	Fill in the blanks	Fill in the blanks	Write the fact
10 x 2 = 20	10 x 2 = 20	10 x 2 = 20	10 x 2 = 20	10 x 2 = 20
10 x 7 = 70	10 x 7 = 70	10 x 7 = 70	10 x 7 = 70	10 x 7 = 70
10 x 3 = 30	10 x 3 = 30	10 x 3 = 30	10 x 3 = 30	10 x 3 = 30
10 x 6 = 60	10 x 6 = 60	10 x 6 = 60	10 x 6 = 60	10 x 6 = 60
10 x 8 = 80	10 x 8 = 80	10 x 8 = 80	10 x 8 = 80	10 x 8 = 80
10 x 10 = 100	10 x 10 = 100	10 x 10 = 100	10 x 10 = 100	10 x 10 = 100
10 x 0 = 0	10 x 0 = 0	10 x 0 = 0	10 x 0 = 0	10 x 0 = 0
10 x 5 = 50	10 x 5 = 50	10 x 5 = 50	10 x 5 = 50	10 x 5 = 50
10 x 9 = 90	10 x 9 = 90	10 x 9 = 90	10 x 9 = 90	10 x 9 = 90
10 x 1 = 10	10 x 1 = 10	10 x 1 = 10	10 x 1 = 10	10 x 1 = 10
10 x 4 = 40	10 x 4 = 40	10 x 4 = 40	10 x 4 = 40	10 x 4 = 40

Page 84

Let's Practice!

Let's make sure you have all your facts down! Answer each multiplication problem.

$$10 \times 3 = 30$$

$$10 \times 8 = 80$$

$$10 \times 6 = 60$$

$$10 \times 10 = 100$$

$$10 \times 2 = 20$$

$$10 \times 5 = 50$$

$$10 \times 7 = 70$$

$$10 \times 0 = 0$$

$$10 \times 1 = 10$$

$$10 \times 4 = 40$$

$$10 \times 9 = 90$$

Page 85

I can multiply by 10: TOP or BOTTOM

Name: _____ Date: _____
Goal: ____ problems in ____ seconds/minutes

10 x 4 = 40	10 x 3 = 30	10 x 8 = 80	10 x 4 = 40	10 x 10 = 100	10 x 2 = 20
10 x 5 = 50	10 x 8 = 80	10 x 5 = 50	10 x 2 = 20	10 x 3 = 30	10 x 9 = 90
10 x 7 = 70	10 x 8 = 80	10 x 4 = 40	10 x 9 = 90	10 x 2 = 20	10 x 0 = 0
10 x 1 = 10	10 x 0 = 0	10 x 5 = 50	10 x 10 = 100	10 x 3 = 30	10 x 2 = 20
10 x 9 = 90	10 x 4 = 40	10 x 7 = 70	10 x 5 = 50	10 x 2 = 20	10 x 8 = 80
9 x 10 = 90	4 x 10 = 40	7 x 10 = 70	9 x 10 = 90	2 x 10 = 20	0 x 10 = 0
10 x 10 = 100	4 x 10 = 40	7 x 10 = 70	4 x 10 = 40	3 x 10 = 30	2 x 10 = 20
8 x 10 = 80	4 x 10 = 40	7 x 10 = 70	6 x 10 = 60	2 x 10 = 20	0 x 10 = 0
10 x 2 = 20	6 x 10 = 60	4 x 10 = 40	8 x 10 = 80	5 x 10 = 50	9 x 10 = 90
10 x 10 = 100	9 x 10 = 90	2 x 10 = 20	3 x 10 = 30	8 x 10 = 80	5 x 10 = 50

10 Facts Answer Keys

Page 86

I can multiply by 10s: Top AND Bottom

Name: _____ Date: _____
Goal: _____ problems in _____ seconds/minutes

10 × 8 = 80	10 × 10 = 100	3 × 10 = 30	10 × 8 = 80	4 × 10 = 40	10 × 2 = 20
9 × 10 = 90	4 × 10 = 40	10 × 3 = 30	10 × 9 = 90	0 × 10 = 0	10 × 1 = 10
7 × 10 = 70	10 × 7 = 70	10 × 3 = 30	10 × 9 = 90	5 × 10 = 50	2 × 10 = 20
10 × 9 = 90	10 × 3 = 30	2 × 10 = 20	8 × 10 = 80	10 × 9 = 90	10 × 4 = 40
10 × 2 = 20	7 × 10 = 70	10 × 9 = 90	10 × 10 = 100	10 × 0 = 0	4 × 10 = 40
9 × 10 = 90	2 × 10 = 20	10 × 8 = 80	4 × 10 = 40	5 × 10 = 50	10 × 8 = 80
1 × 10 = 10	10 × 8 = 80	10 × 3 = 30	10 × 9 = 90	0 × 10 = 0	2 × 10 = 20
10 × 7 = 70	3 × 10 = 30	7 × 10 = 70	8 × 10 = 80	10 × 9 = 90	2 × 10 = 20
2 × 10 = 20	10 × 10 = 100	5 × 10 = 50	10 × 0 = 0	2 × 10 = 20	10 × 4 = 40
10 × 9 = 90	3 × 10 = 30	4 × 10 = 40	10 × 7 = 70	5 × 10 = 50	10 × 10 = 100

Page 87

REVIEW: 0-10

Name: _____ Date: _____
Goal: _____ problems in _____ seconds/minutes

3 × 10 = 30	7 × 6 = 42	2 × 1 = 2	0 × 8 = 0	7 × 10 = 70	10 × 10 = 100
6 × 7 = 42	3 × 7 = 21	8 × 8 = 64	9 × 2 = 18	10 × 8 = 80	6 × 5 = 30
2 × 4 = 8	5 × 3 = 15	6 × 6 = 36	8 × 3 = 24	10 × 7 = 70	4 × 4 = 16
0 × 9 = 0	3 × 7 = 21	6 × 6 = 36	4 × 2 = 8	9 × 10 = 90	4 × 2 = 8
4 × 6 = 24	10 × 10 = 100	8 × 6 = 48	5 × 2 = 10	4 × 3 = 12	8 × 2 = 16
2 × 2 = 4	5 × 4 = 20	6 × 3 = 18	4 × 9 = 36	10 × 3 = 30	3 × 5 = 15
4 × 6 = 24	7 × 2 = 14	9 × 4 = 36	2 × 8 = 16	9 × 9 = 81	2 × 4 = 8
2 × 3 = 6	6 × 5 = 30	8 × 3 = 24	2 × 6 = 12	7 × 5 = 35	0 × 6 = 0
3 × 10 = 30	8 × 5 = 40	2 × 6 = 12	7 × 5 = 35	7 × 7 = 49	2 × 5 = 10
1 × 3 = 3	4 × 10 = 40	9 × 7 = 63	5 × 6 = 30	2 × 4 = 8	5 × 2 = 10

Page 88

REVIEW: 0-10

Name: _____ Date: _____
Goal: _____ problems in _____ seconds/minutes

3 × 4 = 12	6 × 2 = 12	1 × 8 = 8	9 × 5 = 45	3 × 6 = 18	1 × 0 = 0
10 × 7 = 70	6 × 5 = 30	4 × 7 = 28	8 × 2 = 16	3 × 3 = 9	9 × 9 = 81
7 × 5 = 35	3 × 8 = 24	0 × 5 = 0	3 × 10 = 30	9 × 8 = 72	3 × 3 = 9
4 × 3 = 12	5 × 7 = 25	8 × 3 = 24	2 × 9 = 18	0 × 4 = 0	2 × 3 = 6
3 × 5 = 15	6 × 2 = 12	7 × 7 = 49	8 × 4 = 32	3 × 7 = 21	1 × 9 = 9
0 × 10 = 0	6 × 4 = 24	3 × 6 = 18	7 × 4 = 28	5 × 5 = 25	2 × 9 = 18
4 × 7 = 28	9 × 9 = 81	7 × 4 = 28	2 × 6 = 12	5 × 7 = 35	9 × 4 = 36
4 × 5 = 20	2 × 3 = 12	8 × 8 = 64	4 × 9 = 36	5 × 1 = 3	5 × 3 = 15
7 × 7 = 49	4 × 3 = 12	2 × 8 = 16	9 × 10 = 90	5 × 6 = 30	7 × 3 = 21
8 × 4 = 32	7 × 6 = 42	3 × 6 = 18	5 × 5 = 25	0 × 5 = 0	2 × 2 = 4

Page 89

REVIEW: 0-10

Name: _____ Date: _____
Goal: _____ problems in _____ seconds/minutes

4 × 6 = 24	2 × 7 = 14	10 × 9 = 90	10 × 10 = 100	7 × 4 = 28	2 × 3 = 6
5 × 7 = 35	7 × 2 = 14	8 × 6 = 48	3 × 5 = 15	2 × 9 = 18	6 × 0 = 6
5 × 6 = 30	7 × 2 = 14	8 × 8 = 64	7 × 4 = 28	2 × 9 = 18	10 × 8 = 80
4 × 3 = 12	2 × 6 = 12	8 × 4 = 32	5 × 5 = 25	2 × 7 = 14	6 × 3 = 18
5 × 7 = 35	8 × 8 = 64	6 × 3 = 24	2 × 9 = 18	3 × 7 = 21	6 × 4 = 24
5 × 2 = 18	6 × 5 = 49	3 × 7 = 24	9 × 5 = 5	2 × 7 = 28	7 × 5 = 35
8 × 5 = 10	6 × 7 = 30	8 × 7 = 21	6 × 4 = 45	5 × 3 = 14	3 × 10 = 36
2 × 4 = 40	10 × 10 = 42	8 × 6 = 16	5 × 7 = 24	2 × 8 = 15	9 × 7 = 30
9 × 8 = 8	5 × 5 = 100	6 × 6 = 48	7 × 8 = 35	2 × 0 = 16	3 × 1 = 63
	72	20	36	56	0	3

www.ingramcontent.com/pod-product-compliance
Lightning Source LLC
Chambersburg PA
CBHW082245310526
45795CB00015B/2975